NCRP REPORT No. 91

Recommendations on Limits for Exposure to Ionizing Radiation

Recommendations of the
NATIONAL COUNCIL ON RADIATION PROTECTION AND MEASUREMENTS

Issued June 1, 1987

National Council on Radiation Protection and Measurements
7910 WOODMONT AVENUE / BETHESDA, MD 20814

Library of Congress Cataloging-in-Publication Data

National Council on Radiation Protection and Measurements.
Recommendations on limits for exposure to ionizing radiation.

(NCRP report; no. 91)
"Issued June 1, 1987."
Bibliography: p.
Includes index.
1. Ionizing radiation—Safety measures. 2. Ionizing radiation—Dosage—Standards. I. Title. II. Series.
RA569.N353 1987 362.1′79 87-14070
ISBN 0-913392-89-8

Preface

The National Council on Radiation Protection and Measurements (NCRP) has the responsibility under its charter from the Congress of the United States to:

(1) Collect, analyze, develop and disseminate in the public interest information and recommendations about (a) protection against radiation and (b) radiation measurements, quantities, and units, particularly those concerned with radiation protection;

(2) Provide a means by which organizations concerned with the scientific and related aspects of radiation protection and of radiation quantities, units, and measurements may cooperate for effective utilization of their combined resources, and to stimulate the work of such organizations;

(3) Develop basic concepts about radiation quantities, units, and measurements, about the application of these concepts, and about radiation protection;

(4) Cooperate with the International Commission on Radiological Protection, the Federal Radiation Council, the International Commission on Radiation Units and Measurements, and other national and international organizations, governmental and private, concerned with radiation quantities, units, and measurements and with radiation protection.

Limits for exposure to ionizing radiation constitute a central element of the recommendations specified in the charter.

The last general recommendations of the NCRP on such limits were issued in 1971. At that time, it was not possible to utilize estimates of the risk attributable to radiation exposure as a basis for the limits specified. Subsequent development of our knowledge of radiation risks has made it possible to base radiation exposure limits on risk estimates, and this Report uses that approach.

New recommendations are provided both for occupational and public exposure. The recommendations given are believed to be in concert with current concepts and understanding of radiation risks and their magnitude. However, the NCRP keeps abreast of developments in the field of radiation protection and is prepared to make modifications as knowledge of radiation effects improves.

The International Systemé of Units (SI) is used in this Report followed by the conventional units in parentheses, in accordance with the procedure set forth in NCRP Report No. 82, *SI Units in Radiation Protection and Measurements.*

NCRP Scientific Committee 1 on Basic Radiation Protection Criteria produced a draft report on the possible development of a full risk system and presented it to the Council in 1983. While many favored this approach, it was evident that many other Council members felt that such a system was not ready for adoption. In studying the Council members' responses, Scientific Committee 1 decided to break up the contents of its 1983 draft document into multiple reports.

The Committee decided that the development of these reports would be undertaken by drafting groups which the Committee established. Serving on Scientific Committee 1 at that time (*i.e.*, through 1984) were:

G.W. Casarett, Chairman
University of Rochester Medical Center
Rochester, New York

S. Abrahamson
Department of Zoology
University of Wisconsin
Madison, Wisconsin

W.J. Bair
Battelle Pacific Northwest Laboratories
Richland, Washington

V.P. Bond
Brookhaven National Laboratory
Associated Universities, Inc.
Upton, New York

C.B. Meinhold
Brookhaven National Laboratory
Associated Universities, Inc.
Upton, New York

H.H. Rossi
Radiological Research Accelerator Facility
Nevis Laboratories
Columbia University
Irvington, New York

E.L. Saenger
Radioisotope Laboratory
Cincinnati General Hospital
Cincinnati, Ohio

J.N. Stannard
University of California
San Diego, California

J.B. Storer
Oak Ridge National Laboratory
Oak Ridge, Tennessee

The members of the Drafting Group that prepared the current Report are:

G.W. Casarett, Chairman
University of Rochester Medical Center
Rochester, New York

C.B. Meinhold
Brookhaven National Laboratory
Associated Universities, Inc.
Upton, New York

W.J. Bair
Battelle Pacific Northwest Laboratories
Richland, Washington

W.K. Sinclair
National Council on Radiation Protection and Measurements
Bethesda, Maryland

NCRP Secretariat, **W. M. Beckner**

The Council wishes to express its appreciation to the members of the Committee for their dedicated work over many years and to the Drafting Group for the time and effort devoted to the preparation of this Report.

Bethesda, Maryland
March 11, 1987

Warren K. Sinclair
President, NCRP

Contents

1. Introduction

The National Council on Radiation Protection and Measurements (NCRP) published its last complete set of basic recommendations specifying limits for ionizing radiation in 1971, in NCRP Report No. 39, *Basic Radiation Protection Criteria* (NCRP, 1971a). Although this Report is not intended to be a complete revision of NCRP Report No. 39, in that the background material in that report has not been revised, *the recommendations contained herein replace all of the recommendations given in the 1971 Report.* In addition, this Report contains other basic recommendations first set out in recently published NCRP Reports. Much of the other material in NCRP Report No. 39 has been, and will continue to be, updated in other documents in the NCRP Report Series. Therefore, NCRP Report No. 39 is now superseded.

In NCRP Report No. 39, some estimates of risk were presented, but these were not explicitly utilized in the derivation of recommended dose equivalent limits. The International Commission on Radiological Protection (ICRP), in its Publication 26, presented risk estimates to demonstrate the acceptability of its basic dose equivalent limits (ICRP, 1977a); however, the ICRP did not derive their dose equivalent limits from those risk estimates. Subsequent updating of portions of ICRP Publication 26 have not brought about any closer coupling between their risk estimates and their dose equivalent limits.

The NCRP believes that a logical direction in the evolution of the basis of radiation protection standards from the present system based on dose equivalent used by the NCRP and the ICRP is toward the development of an approach based specifically on risk. In such a system, generally acceptable levels of risk form the basis for radiation protection standards according to estimates of risk per unit of absorbed dose (or other convenient measure) for the various types of radiation. Furthermore, although physical measurements would still be used, the limits themselves would be expressed in terms of risk. However, in view of the current degree of uncertainty in risk estimates, for both external and internal exposure, and the fact that the details of the implementation of such a system have yet to be elaborated, the Council believes that additional review and evaluation are required prior to the introduction of a radiation protection system in which the limitation of exposure is based solely on risk. For the present, therefore, the

Council adopts, in principle, the effective dose equivalent system used by the ICRP, but has modified and updated this approach in several respects, as discussed in this Report.

In developing specific recommendations, attention must be given to the magnitude and uncertainties in risk estimates. The ICRP (ICRP, 1977a) utilized risk estimates that, in the light of more recent information, may be low by an undetermined amount, perhaps a factor of two or more. However, until there is a more complete examination of the entire background of risk estimation, including the reevaluation of the dosimetry of the Japanese atomic bomb survivors, it is difficult to specify just what these values should be. Therefore, the NCRP will use the values of the ICRP (ICRP, 1977a) in this Report. (For additional information see Section 7.) Because the coupling between risks and dose limits is still quite loose, the use of these particular estimates at this time does not critically affect the choice of effective dose equivalent levels for limits or guidance.

Radiation protection systems and recommendations, in addition to being based as much as possible on quantitative data on irradiation and the resulting risk to health, also require value judgments in which experience plays an important role. In addition, human experience with hazards in other aspects of society must also be taken into account. Thus, the level of protection of radiation workers should be comparable with that for workers in other industrial occupations. Similarly, in the radiation protection of the general public, risk should be limited to levels comparable with those experienced in other circumstances to which the public is normally exposed.

Clearly, the improvement of radiation protection is an evolutionary process based on the accumulation and utilization of pertinent knowledge of radiation risks and experience in the management of these and other hazards. The NCRP, in fostering such improvement, keeps aware of developments relevant to the field of radiation protection.

2. Radiation Protection Goals and Philosophy

The goal of radiation protection is to limit the probability of radiation induced diseases in persons exposed to radiation (somatic effects) and in their progeny (genetic effects) to a degree that is reasonable and acceptable in relation to the benefits from the activities that involve such exposure.

2.1 Effects of Concern in Radiation Protection

The serious radiation-induced diseases of concern in radiation protection fall into two general categories: stochastic effects and nonstochastic effects.

For the purposes of this Report, a stochastic effect is defined as one in which the probability of occurrence increases with increasing absorbed dose but the severity in affected individuals does not depend on the magnitude of the absorbed dose. A stochastic effect is an all-or-none response as far as individuals are concerned. A stochastic effect might arise as a result of radiation injury of a single cell or substructure such as a gene and is assumed to have no absolute dose threshold, despite the fact that currently available observations in population samples do not exclude zero effects at low radiation levels. Cancers (solid malignant tumors and leukemia) and genetic effects are regarded as the main stochastic effects or risks to health from exposure to ionizing radiation at low absorbed doses.

A nonstochastic effect of radiation exposure is defined as a somatic effect which increases in severity with increasing absorbed dose in affected individuals, owing to damage to increasing numbers of cells and tissues. Nonstochastic late effects, *e.g.*, diseases characterized by organ atrophy and fibrosis, are basically degenerative, as contrasted with the neoplastic growth characteristic of cancer. In general, considerably larger absorbed doses are required to cause nonstochastic effects to a degree of severity which seriously impairs health, as compared with absorbed doses required for a significant increase in cancer incidence. The incidence of nonstochastic effects in a population may increase with increasing absorbed dose, owing to differences in susceptibility and other contributing causes among individuals in the population. Examples of nonstochastic effects attributable to radiation exposure are lens opacification, blood changes, and a decrease in sperm production in the male.

The question of radiation dose thresholds for nonstochastic somatic effects in tissues or organs is complex, and, in practice, depends heavily on the biological endpoints considered and their precise definition. However, if the nonstochastic endpoints of concern are limited to those that are clinically significant or seriously health-impairing diseases, the thresholds for these endpoints are sufficiently high that limits can be selected so that these effects are avoided. (Conceivably, exposures in the vicinity of the nonstochastic limits could add somewhat to the severity of such diseases in some individuals who develop the diseases largely as a consequence of other causes and who may, therefore, be the highly susceptible individuals.)

Certain clinically significant teratogenic effects of radiation exposure, *i.e.*, structural anomalies or deficits in development or growth, as consequences of irradiation of the embryo or fetus, may have very low radiation dose thresholds during gestational periods that are highly critical for organogenesis and/or growth of the organs involved. Such effects may increase in frequency with absorbed dose, and may also have the nonstochastic characteristic of increasing in severity with absorbed dose. In humans, development and growth of the central nervous system is particularly radiosensitive in this regard over substantial periods of time during gestation (see Section 11). The radiation induced injury of relatively small numbers of growing and differentiating cells is amplified greatly by the process of development and growth.

2.2 Objectives of Radiation Protection

The specific objectives of radiation protection are: (1) to prevent, to the extent practicable, the occurrence of severe radiation-induced nonstochastic diseases by adhering to dose equivalent limits that are below the apparent practical threshold dose equivalent levels; and (2) to limit risk of the stochastic effects, fatal cancer and genetic effects, to a reasonable level in comparison with non-radiation risks and in relation to societal needs, benefits gained and economic factors. These objectives are achieved by applying individual dose equivalent limits for occupational and nonoccupational (general public) exposures.

It is emphasized that for the purposes of radiation protection, a cautious assumption is made, the reliability of which has not been established. This is the assumption that *the dose-risk relationship is strictly proportional (linear) without threshold, throughout the range of dose equivalent and dose equivalent rates of importance in routine radiation protection.* Furthermore, doses and the probability of response (risk) are assumed to accumulate linearly. At higher doses, received acutely, such as in accidents, more complex (non-linear) dose-risk relationships may apply.

Under these assumptions, any selected dose equivalent limit will have an associated level of risk. For this reason, the NCRP fully endorses previous recommendations (FRC, 1960; NCRP, 1971a; ICRP, 1977a) concerning: (1) the need to justify any activity which involves radiation exposure on the basis that the expected benefits exceed the predicted cost (justification), (2) the need to reduce the total radiation detriment from such justifiable activities or practices to as low a level as is reasonably achievable (ALARA), economic and social factors being taken into account, and (3) the need to apply individual effective dose equivalent limits to ensure that the procedures for justification and ALARA do not result in individuals or groups of individuals exceeding levels of acceptable risk. Optimization, as defined by the ICRP in its Publication 37 (ICRP, 1983) is recognized to have the same meaning as ALARA, which is the term that will be used in this document.

This Report is primarily concerned with the second and third items specified above, namely, ALARA and dose equivalent limits. As will be seen in Section 8 of this Report, the effective dose equivalent limit is the upper limit of acceptability and is therefore to be considered a boundary condition, rather than a design criterion. Thus, it is usually inappropriate to design a barrier that only limits the irradiation of the maximally exposed individual to the limit.

In many applications, ALARA is simply the continuation of good radiation protection programs and practices which traditionally have been very effective in keeping the average exposure well below the limits (NCRP, 1978).

Approaches employing quantitative estimates of total radiation detriment and costs of protection have been developed recently by the ICRP (ICRP, 1983). Application of these and other quantitative approaches to the making of decisions for reducing radiation levels to as low as reasonably achievable will be presented in future NCRP Reports.

3. Bases for Occupational Exposure Limits: Comparable Risk

3.1 Introduction

The NCRP believes that the level of protection provided for radiation workers should, as far as possible, be comparable with that in other "safe" industries. Few industries, however, present comparable hazards, in that the special risks of ionizing radiation at low doses—stochastic effects, namely, the induction of cancer and severe genetic effects—are not commonly found in other industries.

One possible basis for comparison would be occupations involving exposure to chemicals, some of which induce cancer. While there are certain similarities in dose-risk relationships, in latencies and other features of risk versus time, there are also some important differences between carcinogenic chemicals and ionizing radiation. A substantial body of information is now developing on the carcinogenic effects of industrial chemicals, see *e.g.*, Schottenfeld and Haas (1978), and the study of these effects offers a better understanding of carcinogenic risks and the underlying mechanisms, as well as practical information on dose-risk relationships. (A report being prepared by the NCRP Task Group on the Comparative Carcinogenicity of Pollutant Chemicals will address this subject.) Information on the risks involved in chemical work, however, is not sufficiently complete, at present, to permit its use as a basis for establishing reasonable levels of safety in radiation work.[1]

As a result, it is necessary to assess the possibility of drawing other comparisons. Ideally, one should compare the total risk for radiation workers with the total risk for workers in other "safe" industries. The total risk for radiation workers includes the risk of accidents in the various radiation occupations and the special risks of radiation exposure. Annual values of the risk of fatal accidents, and also of injuries

[1] In some cases both chemicals and ionizing radiation may be present and then both risks should be recognized.

and illnesses, are well known in a broad range of occupations. Consequently, it would be desirable to be able to compare radiation risks with these known rates. However, only approximate comparisons can be made and exact equivalences are not intended.

Comparisons can be made using several endpoints:

(1) Death: The average rates of fatal cancer and severe genetic disorders due to ionizing radiation in one occupation can be compared with the average accidental death rate in another occupation.

(2) Shortened lifespan: The estimated years of life lost from fatal cancers or severe genetic disorders induced by ionizing radiation can be compared with the years of life lost due to accidental death in the workplace.

(3) Industrial harm: The total direct industrial harm includes time lost due to accidental death and accidental non-fatal injury and illness. It can also be assessed for radiation exposure by the predicted years of life lost due to fatal and nonfatal cancer (*i.e.*, including morbidity) and severe genetic defects (ICRP, 1977b; ICRP, 1985a).

The first comparison cited above is the simplest since it equates a measured accidental death rate, which is presumably predictive of future rates, with a predicted rate of fatal radiogenic cancer and severe genetic effects derived from risk estimation. (Radiogenic cancer risk estimation is based on a measured rate of excess fatal cancer in humans, but estimates of serious radiogenic genetic effects in humans are based only on animal data.) For genetic effects, the first two generations of the worker's offspring were chosen by the ICRP (ICRP, 1977a) to reflect the pain and suffering that the worker would directly experience if his or her children, or grandchildren, were victims of severe genetic disorders. It is recognized, of course, that the genetic effect for all future generations (at equilibrium) should be accounted for in assessing the total detriment; this amounts to a factor of about two greater than the harm expressed in the first two generations.

The second comparison cited above can be relatively exact because all that is needed to estimate years of life lost is, on the one hand, the fatal industrial accident rates and the average age of workers at the time of accident, and, on the other hand, the radiogenic risk of fatal cancer or severe genetic effects, the average latency period between exposure and death, and the mean age of the radiation work force.

The third and more complex comparison additionally allows for time lost due to injuries and illness and for morbidity, on the one hand, and for the morbidity due to radiation induced nonfatal cancers and severe genetic effects, on the other. These elements involve considerations not easily quantified or compared, partly because of a lack of available data, but important progress is being made in this area (ICRP, 1985a).

It may be noted that on the basis of average years of life lost, fatal accidents tend to shorten affected workers' lives more than do fatal radiogenic cancers (because of the latency of the latter) by a factor of two or more (ICRP, 1977b).

In none of the above are modifications in the quality of life resulting from these effects taken into account. Clearly it would be valuable to make the comparisons more exact by comparing years of quality adjusted life lost, rather than simply years of life lost for accidental circumstances versus those for radiation induced cancers. Again, progress is being made in this field, but the techniques are presently beyond the scope of this text (see, for example, Pliskin *et al.*, 1980).

3.2 Annual Fatality Rates from Accidents in Various Industries

The National Safety Council (NSC) has reported that the *average annual rate* for accidental death in all U.S. industry in the year 1984 (which is typical of recent years) was about 1.1 per 10,000 workers (1.1×10^{-4}) with a range of 0.5×10^{-4} for trade to 5×10^{-4} for agriculture (0.5 to 5 per 10,000 workers) and 6×10^{-4} for mining and quarrying (NSC, 1985). As can be seen in Table 3.1, the mean annual value of approximately 1 in 10,000 workers (10^{-4}) also separates the nominally safe industries (manufacturing, trade, service, and government) from the less safe industries (public utilities and transportation, construction, mining, and agriculture). *Therefore, "safe" industries are defined here as those having an associated annual fatality accident rate of 1 or less per 10,000 workers, i.e., an average annual risk of 10^{-4}.* It is important to note that these figures have been decreasing rather steadily, at a rate of about two percent per year.

For perspective, it should also be noted that traveling to and from work in the United States, which is mostly by automobile, has an average annual risk of mortality of about 10^{-4} (Sinclair, 1985a), about the same as the accident rate currently existing for all industries.

The risks among subgroups within an industry vary widely, and in one assessment (ICRP, 1985a) the ratio of subgroup fatal accident rates to the mean varied by up to a factor of 10 or more in either direction, but averaged about a factor of five. Thus, broadly speaking, the ratio of the maximum fatal accident rate to the average rate is

TABLE 3.1—*Annual fatality rates from accidents in different occupations*[a] (NSC, 1985)

Occupation	Number of workers $\times 10^3$	Annual fatal accident rate (per 10,000 workers)
Trade	24,000	0.5
Manufacturing	19,900	0.6
Service	28,900	0.7
Government	15,900	0.9
Transportation and utilities	5,500	2.7
Construction	5,700	3.9
Agriculture	3,400	4.6
Mining, quarrying	1,000	6.0
All industries (U.S.)	104,300	1.1

[a] Certain occupations have higher annual fatal accident rates than those given here. See Section 7.

about five, and for the minimum rate to the average it is about one-fifth.

The ratios of average annual doses for various subgroups of radiation workers to the mean for all workers are not well documented, but obviously may range up to about 20 since the annual limit is about 20 times the average exposure for all workers. These ratios of subgroup exposures are of the same order as those for subgroup accident rates in other industries.

Because of the difference between radiation induced fatal cancers and fatal accidents and the existence of subgroups with higher and lower risks than the average, comparison between these two is necessarily inexact, but nevertheless provides a useful general basis for deriving occupational exposure limits.

As noted above, the normal risks of occupations involving radiation work form a part of the total risk for these workers. These include accident and illness rates for the different occupations involving radiation, which have various individual rates. For example, accident and illness rates for those involved in radiology and nuclear medicine probably differ from those occurring in industrial, research laboratory and nuclear power uses of radiation. Few of these are known precisely, but those that are known seem comparatively low on the scale of occupational hazards. For example, among U.S. Department of Energy (DOE) workers exposed to ionizing radiation, annual fatal (non-radiation) accident rates of about 0.25×10^{-4} have been reported (DOE, 1984). This compares favorably with the average annual accidental death rate of about 1×10^{-4} in "safe" industries. Data from other occupations are more scant, but some are available from the United Kingdom, where an average annual fatal accident rate of

0.27×10^{-4} for the twenty years 1964–83 has been established for all radiation industries (ICRP, 1985a). These are similar to those reported for DOE workers.

3.3 Criteria for Protection of Radiation Workers

In this Report, the recommendations for protection of radiation workers are based on the following basic criteria: (1) that nonstochastic effects in tissues are essentially avoided at low absorbed doses; (2) that the predicted risk of stochastic effects (*i.e.*, the incidence of fatal cancer among radiation workers and of severe genetic effects in the first two generations of their offspring), together with fatal accident rates for radiation workers, should, on the average, be no greater than the average risk of accidental death among the workers in "safe" industries; and (3) that these risks should be kept as low as reasonably achievable (ALARA) taking into account social and economic factors.

From the limited information available, it seems that annual fatal accident rates (non-radiation) may be of the order of 0.25×10^{-4} for occupations involving radiation exposure. For the radiation component, the average annual effective dose equivalent of workers is taken to be 2.3 mSv (0.23 rem) (EPA, 1984), which, utilizing the nominal value of lifetime risk of fatal cancer of 10^{-2} Sv^{-1} (10^{-4} rem^{-1}) (see Section 7) yields, eventually, an annual risk of about 0.25×10^{-4}, comparable with the non-radiation fatal accident rate for occupations involving radiation exposure (Sinclair, 1985a). Thus, radiation workers have ultimately a total annual risk of fatality of about 0.5×10^{-4}, similar to that for the safest of the other groups (see Table 3.1). These, of course, are average values. Even allowing for substantial error in the nominal value of the risk coefficient, uncertainties in the fatal accident rate, and the inexact nature of the comparison, the average annual risk for radiation workers is still unlikely to exceed the annual risk of about 10^{-4} in "safe" industries.

Workers in uranium mines represent a special case, with higher accident rates (as have other miners, see Table 3.1) and higher average radiation exposures than workers in other occupations involving radiation exposure. The annual accidental death rates for uranium miners in Canada, 4.0×10^{-4}, tend to be less than those for non-radiation mining at 5.0×10^{-4} for all hard rock mining in Canada (ICRP, 1985a). Even with a relatively high assumed value for the radiation component, the total risk for uranium miners does not exceed that for hard rock miners (see Table 31 in *Quantitative Basis for Developing a Unified Index of Harm*, ICRP, 1985a).

4. Absorbed Dose, Quality Factor and Dose Equivalent

4.1 Introduction

Radiations differ in their relative biological effectiveness per unit of absorbed dose, and this is accounted for in the dose equivalent, H, which is the product of the absorbed dose, D, and the quality factor, Q, for the radiation in question.[2] The absorbed dose, D, is the energy absorbed per unit mass of irradiated material. The quality factor, Q, is a dimensionless factor selected to account for the difference in biological effectiveness of different types of radiation, within the range of doses of concern in radiation protection activities. Although Q is assumed to be independent of any specific endpoint, it is applicable at low doses for the biological effects of principal concern, *i.e.*, cancer induction and severe genetic defects. Q is only to be used at low absorbed doses in the range of the annual radiation protection limits, and not for the high doses that might occur in accident situations.

The effective quality factor ($\bar{Q}$) is chosen after surveying measured values of the relative biological effectiveness (RBE)[3] of the radiation for a variety of relevant biological effects at low absorbed doses, including those on human material when available. Judgment must be exercised in this choice.

A useful empirical relationship has been established between the quality factor and the linear energy transfer (L_∞) in water of the radiation in question (ICRP, 1977a) and from this, Q values for radiations of specified energy can be determined, or values for effective quality factor ($\bar{Q}$) for particular radiation spectra can be calculated (ICRU, 1970, 1986; NCRP, 1971b). Q can also be defined in terms of the lineal energy (y) of the radiation in soft tissue (ICRU, 1986).

[2] The dimensionless factor N once used in this equation is deleted, as recommended by the Joint Task Group of the ICRP-ICRU (ICRU, 1986).

[3] Note that RBE is defined as the ratio of the absorbed dose of a "reference" radiation to that of the test radiation required to produce the same degree of biological effect, determined, as far as possible, from full dose-response information (ICRP-ICRU, 1963).

4.2 Recommended Values of $\bar{Q}$

For many purposes, it is permissible to use approximate values (*i.e.*, values that do not underestimate the actual value) of $\bar{Q}$ for well known radiations, and the approximate values of $\bar{Q}$ recommended by the NCRP for this purpose are listed in Table 4.1. However, such approximate values should not be used in cases where greater precision is required and more exact values of $\bar{Q}$ can be determined from known energy spectra.

As indicated in Table 4.1, the $\bar{Q}$ values for low-LET radiations (x rays, γ rays, electrons) are all designated as unity, even though, at low doses, differences in RBE between them have been identified (ICRU, 1986). High-LET radiations (protons, neutrons, alpha particles and heavy ions) of unknown energy are all assigned a $\bar{Q}$ value of 20. The $\bar{Q}$ for thermal neutrons is specified at an intermediate value of five.[4]

The designations in Table 4.1 include affirmation of a $\bar{Q}$ of 20 for alpha particles of unknown energy, in agreement with the range of 1 to 20 given earlier by NCRP (NCRP, 1971a) and with ICRP (ICRP, 1977a) and also include an increase in $\bar{Q}$ for neutrons and protons to 20 from the former value of 10.

4.3 Basis of Recommended Values of $\bar{Q}$

The NCRP issued a statement in 1980 (NCRP, 1980) drawing attention to the possibility that protection for neutrons might not be comparable with that for other radiations. In 1985, the ICRP (ICRP, 1985b) issued a statement which includes the following:

> "The information now available on the relative biological effectiveness (RBE) for neutrons for a variety of cellular effects *in vitro*, and for life-shortening in the mouse, is being reviewed by the Commission. The implications of this information will be considered as part of a larger review of the recommendations to be undertaken by the Commission over the next four years or so. Meanwhile, in the case of neutrons, the Commission recommends an increase in Q by a factor of 2. The permitted approximation for $\bar{Q}$ for fast neutrons thus changes from 10 to 20. These changes relate only to neutrons, and no other changes in Q are recommended at this time.

[4] The "old" value was 2.3, but this was recently adjusted to 3.5 prior to the ICRP Statement of 1985 (ICRP, 1985b). The ICRP statement suggests that the 3.5 be multiplied by two. However, until more definitive information is available the NCRP recommends a value of five.

TABLE 4.1—*Recommended Values of $\bar{Q}$ for various types of radiation*

Type of radiation	Approximate value of $\bar{Q}$
X rays, γ rays, β particles and electrons	1
Thermal neutrons	5
Neutrons (other than thermal), protons[a], alpha particles and multiple-charged particles, of unkown energy[b]	20

[a] In circumstances where the human body is irradiated directly by high energy protons, the RBE is likely to be similar to that of low-LET radiations, and therefore a $\bar{Q}$ of about unity would be appropriate for that case. Only low energy protons (*e.g.*, like those generated in tissue by fast neutrons) can be expected to require the high value of $\bar{Q}$ listed here.

[b] If energies are known, $\bar{Q}$ values can be obtained for neutrons by the procedures and values given in NCRP Report No. 38 (NCRP, 1971b) but multiplied by two. For other particles of known energy, see ICRP Publication 21 (ICRP, 1973), but again the values would have to be multiplied by two.

The recent action on neutrons by the ICRP, and the recommendations given here by the NCRP, are based on the growing evidence, from studies in many biological systems *at absorbed doses in the range of 10 to 100 mGy (1 to 10 rad)*, that the relative biological effectiveness of neutrons is greater than that represented by the former quality factor of 10. The experimental evidence is summarized in a report of the ICRU (ICRU, 1986) on the work of a joint ICRP-ICRU Task Group, a forthcoming Report of the NCRP (being developed by NCRP Scientific Committee 40 on Biological Aspects of Radiation Protection Criteria) and in recent papers, such as those of Broerse *et al.*, 1985; Dennis and Dunster, 1985; and Sinclair, 1983, 1985b. By way of example, a summary table of RBE_M (maximum values of RBE at low doses) for neutrons and γ rays, taken from the ICRP-ICRU Task Group Report (ICRU, 1986), is reproduced here as Table 4.2. Similar values, covering at least as broad a range, are given in the 1986 draft of the report being prepared by NCRP Scientific Committee 40, which constitutes a separate evaluation of the data. Some experimental evidence also indicates that there are comparatively few differences in the effectiveness of a variety of types of high-LET radiation such as alpha particles, neutrons and multiply-charged particles and it is therefore appropriate that the same value of Q be applied to each of them.

The $\bar{Q}$ for neutrons (other than thermal) listed in Table 4.1 is for the case when the neutron energy[5] is not known and therefore the greatest effectiveness is assumed. If energies are known, $\bar{Q}$ values can

[5] This includes neutrons of all energies other than thermal neutrons. It includes neutrons in the optimum energy range (most biologically effective) of 0.3 to 0.8 MeV and fission neutrons.

TABLE 4.2—*RBE_M*[a] *for fission (or optimum energy)*[b] *neutrons versus fractionated γ rays*[c]

Type of effect	RBE_M
Tumor induction	≈3–≈200
Life shortening	15–45
Transformation	35–70
Cytogenetic studies	40–50
Genetic endpoints in mammalian systems	10–45
Other endpoints	
Lens opacification	25–200
Micronucleus assay	6–60
Testes weight loss	5–20

[a] "It is convenient for purposes of discussion to denote values of RBE at minimal doses by a special symbol which is here denoted RBE_M. RBE_M is thus the ratio of slopes of the dose effect curves at zero dose" (ICRP-ICRU, 1963). Relatively few determinations of RBE_M have been possible until the last few years when more and more measurements at sufficiently low doses have been made. It is on such measurements that the range of values in this table are based.

[b] "Optimum energy" is the neutron energy for which the relative biological effect is greatest.

[c] When the exposures were not actually fractionated, an estimated risk coefficient for fractionated γ rays was made from risk coefficients obtained under other conditions.

be obtained for neutrons by the procedures and values given in NCRP Report No. 38 (NCRP, 1971b) but multiplied by two. For other particles of known energy, see ICRP Publication 21 (ICRP, 1973), but again the values would have to be multiplied by two.

Values of Q as a function of neutron energy are also given by ICRU (ICRU, 1986), and these values differ slightly from those recommended above. The ICRU (ICRU, 1986) also recognizes values other than one for photons as a function of energy, and considers it permissible to use these when the energy is precisely known. Hard γ rays (>0.5 MeV), for example, are less effective than orthovoltage x rays at low doses by about a factor of two. However, as noted in Table 4.1, the NCRP does not recommend recognition of this difference for protection purposes at this time.

It should be noted that when the human body is irradiated, the spectrum of charged particles produced at a depth in the body differs from that at the surface. Thus, for precise dosimetry, the Q value appropriate to the particle energies at a given location within the body should be estimated as indicated in NCRP Report No. 38 (NCRP, 1971b). While the increase in $\bar{Q}$ for neutrons indicated in Table 4.1 suggests that dose equivalents for neutron exposure may have been underestimated in the past, the method of assessing individual neutron exposure usually overestimates the effective dose equivalent, as already noted in the NCRP statement (NCRP, 1980).

This important subject is presently at a transition stage. Additional studies are needed to improve our understanding of the effects of different radiations at low doses and these will be kept under review by the NCRP.

5. Effective Dose Equivalent

When the whole body is uniformly irradiated, the probability of the occurrence of cancer and genetic effects is assumed to be proportional to the dose equivalent to the whole body, H_{wb}, and the risk may be represented by a single value. However, truly uniform whole body exposures are rare. Non-uniform or partial body exposures usually occur when irradiation is from radionuclides deposited in various tissues and organs. (Even with external irradiation there are differences arising with depth in the body and orientation of the body in the generally non-uniform radiation field.) In some of these cases, the dose equivalents for various tissues, from a single exposure, can differ markedly. Tissues also vary in their sensitivity to radiation-induced cancer. The effective dose equivalent concept, first developed by Scientific Committee M-4 of the NCRP (NCRP SC M-4, 1967) and adopted by the ICRP (ICRP, 1969, 1977a, 1978, 1979, 1980) provides a means for handling these non-uniform irradiation situations and is here adopted by the NCRP.

The effective dose equivalent (H_E), is the sum of the weighted dose equivalents for irradiated tissues or organs. It takes into account the different mortality risks from cancer and the risk of severe hereditary effects in the first two generations associated with irradiation of different organs and tissues. The principle that the risks for all stochastic effects should be the same whether the whole body is irradiated uniformly or non-uniformly is met if:

$$H_E = \sum_T w_T H_T = H_{wb} \tag{5.1}$$

where w_T is the weighting factor representing the proportionate risk (stochastic) of tissue T, when the whole body is irradiated uniformly and H_T is the mean dose equivalent received by tissue T. Weighting factors (w_T) recommended by the ICRP (ICRP, 1977a) are adopted by the NCRP for the purposes of the recommendations in this Report. These weighting factors and the risk coefficients from which they were derived are given in Table 5.1. The system of dose limitation employing these weighting factors specifies that for annual exposures:

$$\sum_T w_T H_T \quad \text{or} \quad H_{wb} \leq H_{E,L}, \tag{5.2}$$

where $H_{E,L}$ is the recommended annual effective dose equivalent limit.

TABLE 5.1—*Recommended values of the weighting factors, w_T, for calculating effective dose equivalent and the risk coefficients from which they were derived*
(Values from ICRP, 1977a)[a]

Tissue (T)	Risk coefficient	w_T
Gonads	40×10^{-4} Sv^{-1} (40×10^{-6} rem^{-1})	0.25
Breast	25×10^{-4} Sv^{-1} (25×10^{-6} rem^{-1})	0.15
Red bone marrow	20×10^{-4} Sv^{-1} (20×10^{-6} rem^{-1})	0.12
Lung	20×10^{-4} Sv^{-1} (20×10^{-6} rem^{-1})	0.12
Thyroid	5×10^{-4} Sv^{-1} (5×10^{-6} rem^{-1})	0.03
Bone surfaces	5×10^{-4} Sv^{-1} (5×10^{-6} rem^{-1})	0.03
Remainder[b]	50×10^{-4} Sv^{-1} (50×10^{-6} rem^{-1})	0.30
Total[c]	165×10^{-4} Sv^{-1} (165×10^{-6} rem^{-1})	1.00

[a] Possible future modifications to the values in this table are discussed in Section 7.

[b] A w_T of 0.06 is to be assigned to each of the five remainder tissues receiving the highest dose equivalents and the other remainder tissues are to be neglected. (When the gastrointestinal tract is irradiated, the stomach, small intestine, upper large intestine and lower large intestine are to be treated as four separate organs and each may therefore be included in the five remainder tissues depending on the magnitude of the dose equivalent they receive when compared to the dose equivalent received by other remainder tissues and organs.)

[c] The total for somatic risk alone is 125×10^{-4} Sv^{-1} (125×10^{-6} rem^{-1}) which for radiation protection purposes is often rounded to a nominal value of 1×10^{-2} Sv^{-1} (1×10^{-4} rem). Genetic risk is 40×10^{-4} Sv^{-1} (40×10^{-6} rem^{-1}).

The NCRP accepts the ICRP recommendations that exclude extremities (hands and forearms, feet, ankles), skin and lens of the eye from the "remainder" tissues for assessment of effective dose equivalent for individuals (ICRP, 1978). The limits for these tissues, except for skin, are the nonstochastic limits (see Sections 9 and 22). However, for assessment of detriment resulting from exposure of skin, a risk factor of 10^{-4} Sv^{-1} (10^{-6} rem^{-1}) and therefore a weighting factor of 0.01 is to be applied to the mean dose equivalent over the entire skin surface. No adjustment of other weighting factors to maintain a total of one is recommended; however, the NCRP accepts, for the present, the current ICRP recommendation of adding the product of the mean skin dose equivalent and the weighting factor of 0.01 to the effective dose equivalent and that this product should be specified when it is used. The NCRP believes this subject requires further review.

6. Committed Dose Equivalent, Annual Limit on Intake and Derived Air Concentrations

6.1 Committed Dose Equivalent

The absorbed dose from external irradiation is received at the time the tissue is exposed to the radiation. However, for irradiation from internally deposited radionuclides, the absorbed dose will be distributed temporally depending on the effective half-life[6] of the radionuclide involved. The absorbed dose rate from internally deposited radionuclides depends upon the amount of radioactive material (activity) in the body. The distribution of the absorbed dose within the tissues of the body will vary with the route of entry, the chemical and physical properties of the radionuclide, the type and energy of the radiations emitted, the sites of deposition in tissues, the mass and density of the tissues, the rates of deposition and clearance from the sites of deposition, the rates of redistribution and the rates of excretion from the body. To take into consideration the varying time distributions of radiation dose, the ICRP defined the committed dose equivalent in a particular organ or tissue as the total dose equivalent averaged throughout that organ or tissue in the 50 years after intake of a radionuclide into the body. The committed effective dose equivalent is assumed to be a measure of the risk that will result from that intake. [The parameters affecting the committed dose equivalent and limitations in the utilization of the committed dose equivalent are discussed in NCRP Report No. 84 (NCRP, 1985).]

The committed dose equivalent, H_{50}, is the time integral of the dose equivalent rate in a specific tissue following intake of a radionuclide into the body. The formal definition of committed dose equivalent is:

$$H_{\mathrm{T},50} = \int_{t_0}^{t_0+50y} \dot{H}_{\mathrm{T}}\, dt \tag{6.1}$$

for a single intake of a radionuclide at time t_0, where $\dot{H}_{\mathrm{T}}$ is the relevant dose-equivalent rate in an organ or tissue (T) at time t.

[6] $T_E = T_B T_P/(T_B + T_P)$, where T_E is effective half-life, T_B is biological half-life and T_P is physical half-life of the radionuclide.

For radionuclides with approximate effective half-lives ranging up to about three months, the committed dose equivalent is approximately equal to the annual dose equivalent for the year of intake. For radionuclides with a long effective half-life, the committed dose equivalent is greater than the annual dose equivalent because it reflects the dose that will be delivered in the future as well as that delivered during the year of intake. For radionuclides with a very long effective half-life, neither the total dose nor a full expression of the risk can be expected to be manifested during the worker's lifetime. For this reason, the committed dose equivalent from the life-long intake of some radionuclides of long effective half-life will overestimate by about a factor of two, or more, the lifetime risk that can actually be expressed.

The NCRP adopts the committed dose equivalent for radiation protection planning, such as in the design of facilities and development of manufacturing processes, research protocols, etc.; for the demonstration of compliance with those plans; and for use in calculating annual limits of intake (ALIs) and derived air concentrations (DACs). The committed effective dose equivalent may also be used to provide an estimate of the lifetime risk from a given intake and, in this respect, its use is the same as the effective dose equivalent for external irradiation, with the limitation given above. However, it does not constitute a sufficient basis in itself for the evaluation of the potential health effects of radiation exposures in individuals. Such evaluations should be based on estimates of actual absorbed dose, derived from the models given in ICRP Publication 30 (ICRP, 1979), for the period of exposure appropriate to the individual case. For greater accuracy, the acceptable practice is to base these estimates on physiological and exposure information specific to the individual rather than on ICRP models which are for Reference Man (ICRP, 1975).

The committed effective dose equivalent ($H_{E,50}$) for each internally deposited radionuclide is calculated by summing the products of the committed dose equivalents for each tissue irradiated and the appropriate weighting factor,

$$H_{E,50} = \sum_T w_T H_{T,50} \tag{6.2}$$

6.2 Annual Limit on Intake: Occupational

The Annual Limit on Intake (ALI) is the maximum quantity of a radionuclide that can be taken into the body based on ICRP Reference Man (ICRP, 1975) each year without the committed effective dose equivalent being in excess of the annual effective dose equivalent limit (see Section 8) or the committed dose equivalent to any tissue being in excess of the nonstochastic limit (see Section 9). It is a secondary limit. As such, it provides reasonable assurance that, if it is not exceeded, $H_{E,50}$ will not exceed the annual effective dose equivalent limit.

For the purposes of these recommendations, the ALIs given in ICRP Publication 30 (ICRP, 1979) are adopted. These ALIs are for individual radionuclides. The NCRP has reviewed and agrees generally with the dosimetric, metabolic and excretion models and methods currently used by the ICRP in calculating ALIs (NCRP, 1985). Work on these models is continuing and is likely to lead to some revisions of these ALIs in the future.

6.3 Derived Air Concentrations

The ICRP (ICRP, 1979) introduced the Derived Air Concentration (DAC) for planning and to assist in controlling exposures in the workplace to the limit defined by the ALI. The DAC is that concentration of a radionuclide in air that if breathed by Reference Man, inspiring 0.02 m^3 per minute for a working year, would result in an intake of 1 ALI. Thus, the DAC is determined by dividing the ALI by 40 hours per week, 50 weeks per year, 60 minutes per hour and 0.02 m^3 per minute.

$$\mathrm{DAC} = \frac{\mathrm{ALI}}{40\,\mathrm{h\,wk}^{-1} \times 50\,\mathrm{wk\,y}^{-1} \times 60\,\mathrm{min\,h}^{-1} \times 0.02\,\mathrm{m}^3\,\mathrm{min}^{-1}} \tag{8.1}$$

DACs for individual radionuclides published in ICRP Publication 30 (ICRP, 1979) are adopted here. The NCRP agrees with the ICRP in emphasizing that the ALI is a secondary limit, whereas the DAC is only a device for controlling exposures in the workplace to the ALI. Exposure to radionuclide concentrations exceeding the DAC are primarily an indication that a potential for exposure exceeding the ALI exists and that the workplace may not be properly controlled. Since the values for DACs apply to individual radionuclides, they should be reduced appropriately when two or more radionuclides are involved.

DACs calculated for workers cannot, of course, be used to control exposures of members of the public. Differences in factors such as applicable dose equivalent limits, duration of exposure, breathing rate, body size, metabolism and transfer factors would obviate such use (ICRP, 1984a). Further, exposures via other environmental pathways would have to be considered—food and water, for example. On the other hand, derived concentrations of radionuclides in water, for example, could be calculated, when needed, in a manner similar to DACs.

7. Risk Estimates for Radiation Exposure

Estimates of the risks of somatic and genetic effects of irradiation have been formulated by some individuals and by specific organizations such as the United Nations Scientific Committee on the Effects of Atomic Radiation (UNSCEAR), the National Academy of Sciences/National Research Council (NAS/NRC) and the ICRP (Lewis, 1957; UNSCEAR, 1972, 1977; NAS/NRC, 1972, 1980; ICRP, 1977a). The ICRP utilized, for its 1977 recommendations, the values given in Table 5.1. These estimates are similar to those given by UNSCEAR (UNSCEAR, 1977) and supported, in general terms, by the report of the BEIR Committee of the National Academy of Sciences/National Research Council (NAS/NRC, 1980).

The risk estimates given in Table 5.1 include an assumption of full expression of the cancer risk (which is unlikely to occur from exposures received late in life) and an assumption of a population distribution over all ages and both sexes, and it includes severe genetic effects for the first two generations. The somatic component of this total is 1.25 $\times 10^{-2}$ Sv^{-1} (1.25 $\times 10^{-4}$ rem^{-1}) which is an average for both sexes and a normal population age distribution of 18 to 60 years. This is often rounded for protection purposes to a nominal value of 1 $\times 10^{-2}$ Sv^{-1} (1 $\times 10^{-4}$ rem^{-1}). The genetic component of the risk considered to be of significance to individuals are the effects in the first two generations which are estimated to amount to 0.4 $\times 10^{-2}$ Sv^{-1} (0.4 $\times 10^{-4}$ rem^{-1}) (see Table 5.1).

A number of changes in risk estimates have taken place since the last UNSCEAR and ICRP evaluations. These include the addition of further data from the studies of the Japanese survivors (Kato and Schull, 1982), additional data on breast cancer, and a greater appreciation of the probable validity of relative risk projection models. Some of these changes are outlined in the *Report of the National Institutes of Health Ad Hoc Working Group to Develop Radioepidemiological Tables* (NIH, 1985), for which risk was developed as a function of time after exposure. Lifetime risk estimates, using the same updated risk input, have recently been developed for use in a report being prepared by NCRP Scientific Committee 75 concerned with radiation exposures in space activities. Overall, when these new values of risk are averaged over an entire work force of mixed ages and sexes, the nominal value of lifetime risk is likely to exceed 10^{-2} Sv^{-1} (10^{-4} rem^{-1}) by perhaps a factor of two or more. Furthermore, estimates of genetic risk have increased since 1980 and current estimates would also be higher by a factor of about two.

As noted earlier, the NCRP does not propose to use this new information for the purpose of developing the dose limits at this time because: (1) risk estimates are likely to undergo substantial further re-evaluation in the near future because of the completion of the revisions in the dosimetry of the Japanese survivors, expected in 1987, and the addition of yet another cycle of data (1979–82) from the Japanese studies and perhaps a third cycle 1983–85; (2) risk estimates are subject to uncertainties of perhaps up to a factor of about five; and (3) the relationship between limits and risk estimates is not precise. The changes likely to result from the new studies may require revisions in our appreciation of the best values for risk estimates and possibly in weighting factors (w_T), but it is impossible to say at the present time just what these values will be. In the meantime, we will continue to use a nominal lifetime somatic risk of 10^{-2} Sv^{-1} (10^{-4} rem^{-1}) for adults, recognizing that it is subject to change and will probably increase, perhaps substantially.

8. Effective Dose Equivalent Limit: Occupational[7]

For the purpose of deriving the annual effective dose equivalent limit for occupational exposure in this Report, a uniform whole body dose equivalent of 10 mSv (1 rem) is assumed to result in an average nominal lifetime risk of 10^{-4} for fatal stochastic effects for adult workers (see Section 7).

A review of the existing data (EPA, 1984) on occupational exposure reveals that the average annual dose equivalent for monitored workers, with measurable exposures, was approximately 2.3 mSv (0.23 rem) in 1980. This is approximately one-twentieth of the NCRP prospective annual whole body exposure limit of 50 mSv (5 rem) (NCRP, 1971a).

From these occupational exposure data, it would appear that the limits recommended in 1971 (NCRP, 1971a) are adequate to ensure that the risks experienced by radiation workers are comparable to those in "safe" industries (see Section 3). *The NCRP therefore recommends an annual effective dose equivalent limit of 50 mSv (5 rem).* However, in view of (1) the likelihood that the risk estimates will increase over the nominal value used here (perhaps by a factor of two or more), (2) the demonstrated tendency for safe industries to become safer (by a factor of about two every 25 years), and (3) the need to allow for the conventional non-radiation accident mortality risk in the occupations of the radiation workers, the NCRP believes that the recommendations given in its Report No. 39 (NCRP, 1971a) should be strengthened.

First, the NCRP recommends that the use of the cumulative limit, i.e., (Age-18) × 50 mSv [(Age-18) × 5 rem] be discontinued.

Second, the NCRP emphasizes the upper boundary nature of the 50 mSv (5 rem) annual effective dose equivalent limit for routine operations. It is the boundary condition for radiation protection. It is only when the cost of further dose reduction is truly unreasonable that the limit should be approached. It should, of course, be recognized that this recommendation is for routine operations and does not apply to special circumstances such as emergency situations or space activities.

[7] Nonstochastic limits are discussed in Section 9 and a summary of all recommendations on limits is provided in Section 22.

It should be pointed out that a group of workers whose average annual exposure substantially exceeds 10 mSv (1 rem) over their working life would eventually have an average annual excess risk from radiation alone greater than the fatal accident rate for the average worker in safe industries. These infrequently occurring groups of individuals would have an average excess risk level in the upper range of that of specific occupational groups within safe industries or similar to the average excess risk in the more hazardous occupations noted earlier, namely, public utilities and transportation, construction, agriculture and mining. Certain individuals, such as deep-sea fishermen, steeplejacks, loggers and test pilots, work in even more hazardous occupations.

Based on the foregoing considerations and its desire to keep the lifetime risk of fatal stochastic effects to radiation workers below one in one hundred, the Council believes that a third element in strengthening the recommendations is required, and therefore, provides the following guidance: *the community of radiation users is encouraged to control their operations in the workplace in such a manner as to ensure, in effect, that the numerical value of the individual worker's lifetime effective dose equivalent in tens of mSv (rem) does not exceed the value of his or her age in years.* Clearly, this is not intended to permit exposures in excess of the annual limit of 50 mSv (5 rem), nor is it intended to suggest that it is acceptable that younger workers be allowed higher annual exposures than older workers.

The Council believes that such guidance on workplace control is preferable to an age-related lifetime limit which might have undesirable side effects, such as: (1) young workers (with greater risk per unit effective dose equivalent) receiving the greater exposure; (2) the adoption of a more restrictive annual limit which could lead to a larger number of workers being exposed; (3) forcing older highly skilled workers (with lower risk per unit dose equivalent) into occupations with just as high or higher associated risks; and (4) being perceived by some to be an unwarranted infringement on personal liberty. The guidance suggested here is intended to avoid these problems.

In the control of occupational exposure, the application of the dose limits specified here are not sufficient in themselves. As explained earlier, radiation protection philosophy also involves keeping workers' exposures as low as reasonably achievable, economic and social factors being taken into account. Keeping exposures as low as reasonably achievable will generally result in levels substantially below the dose limits (see, for example, Section 14). Further information on the application of ALARA principles is available in ICRP Publication 37 (ICRP, 1983).

9. Nonstochastic Effects Limits: Occupational

The nonstochastic effects of concern are degenerative effects severe enough to be clinically significant and health-impairing diseases (see Section 2).

The weighting factors for the stochastic effects of concern (cancer and genetic effects) would permit very large annual dose equivalents to organs having small weighting factors if such organs were irradiated alone (see Section 5). Therefore, to avoid serious nonstochastic effects that might be produced by high doses under such circumstances, an additional restriction on exposure is recommended.

This is accomplished by limiting individual organs to annual dose equivalents low enough to assure that, in practice, the dose threshold values (ICRP, 1984b; UNSCEAR, 1982) for the nonstochastic effects of concern in various organs and tissues are not likely to be exceeded.

The dose equivalent limits for nonstochastic effects in various organs are intended only to constrain exposures that would not be constrained within these limits by the stochastic effects limits. Thus, in the case of organs with relatively small weighting factors, the nonstochastic effects limits will be overriding, while in organs with relatively large weighting factors the stochastic effects limits will prevail.

For avoidance of nonstochastic effects, *the following annual dose equivalent limits are recommended for the occupational case; 150 mSv (15 rem) for the crystalline lens of the eye, and 500 mSv (50 rem) for all other tissues or organs, including the skin and extremities*. These limits apply whether an individual tissue or organ is exposed selectively or together with other tissues or organs.[8] These recommendations agree with those of the ICRP (ICRP, 1977a) for most organs and for the lens of the eye (ICRP, 1980). The adequacy of the recommended dose

[8] Owing to their susceptibility to nonstochastic radiation damage, it is conceivable that annual irradiations of the active bone marrow at or somewhat below the 0.5 Sv (50 rem) occupational limit for nonstochastic effects, or of the gonads below that limit, could cause nonstochastic damage. In practice, selective irradiation of these organs year after year at such a high dose level is most improbable. Should this situation arise other limits would have to be employed.

equivalent limits for nonstochastic effects is discussed in a recent report of the ICRP (ICRP, 1984b).

In practice, the average annual recorded exposures received by radiation workers is only a small fraction of the effective dose equivalent limit for stochastic effects, generally a tenth or less (EPA, 1984). In addition, workers seldom receive exposures approaching the effective dose equivalent limit repeatedly year after year. Thus, in practice for the occupational case, it is reasonable to expect lifetime effective dose equivalents no greater than age (years) times 10 mSv (1 rem) *i.e.*, mean dose equivalents for organs no greater than (age $\times$ 10 mSv)/w_T [(age $\times$ 1 rem)/w_T]. Also, selective occupational exposure of individual tissues or organs occurs rarely, except in the case of some internally deposited radionuclides; thus, compliance with the stochastic limit assures that the exposure of selected tissues or organs will usually be much less than the nonstochastic limits.

On the basis of the practical considerations discussed above, it can be inferred that the recommended dose equivalent limits provide reasonable safety from nonstochastic health effects for all tissues or organs.

The ALIs given in ICRP Publication 30 (ICRP, 1979) are limited by the recommended nonstochastic limits as well as the stochastic limits.

10. External Plus Internal Exposure: Occupational

To ensure that the annual effective dose equivalent limit ($H_{E,L}$) from all sources is not exceeded, the contributions of internal and external exposure must be summed.

$$H_{E,ext} + H_{E,int} \leq H_{E,L} = 50 \text{ mSv (5 rem)}, \qquad (10.1)$$

where $H_{E,ext}$ and $H_{E,int}$ are the effective dose equivalents from external and internal irradiation, respectively. This can be expressed in secondary quantities:

$$\frac{H_{E,ext}}{H_{E,L}} + \sum_{j,k,l} \frac{I_{j,k,l \ldots}}{ALI_{j,k,l \ldots}} \leq 1, \qquad (10.2)^{9}$$

where $I_{j,k,l \ldots}$ is the annual intake of radionuclides j,k,l..., and $ALI_{j,k,l \ldots}$ is the annual limits of intake for radionuclides j,k,l...

Since the ALI is calculated on the basis of committed effective dose equivalent, comparison of the annual intake of a radionuclide with the ALI in this expression is an assessment of the committed effective dose equivalent. As indicated in Section 6, the NCRP recommends use of the committed effective dose equivalent for planning and for demonstration of compliance. Evaluation of the possible health consequences of an individual's exposure is more appropriately based upon actual dose equivalents delivered and anticipated (NCRP, 1985). The effective dose equivalents from internal irradiation by individual radionuclides can be estimated from measured air concentrations of the radionuclides and/or *in vivo* counting and bioassay data.

The contributions from internal and external exposures must also be summed in assessing compliance with the limit for nonstochastic effects. The summation of external and internal exposures is also appropriate for planned special exposures and emergencies if both types of exposure can occur.

[9] The summation of the ratios of the annual intake of radionuclides to their ALI is a reasonable approximation of $H_{E,int}$ when the effective half-life of the radionuclide is less than a few months but overestimates $H_{E,int}$ when the effective half-life is very long.

11. Protection of the Embryo-Fetus

The NCRP previously addressed irradiation of the embryo-fetus in NCRP Report No. 53, *Review of NCRP Radiation Dose Limit for Embryo and Fetus in Occupationally-Exposed Women* (NCRP, 1977a). The information that has become available (UNSCEAR, 1986) since the publication of NCRP Report No. 53 has been examined. There appears to be no greater, and probably less, risk of cancer induction than was assumed previously. There may also be greater reason to doubt that radiation exposure was the only or the major cause in observed associations between childhood cancer and prenatal diagnostic irradiation. After irradiation *in utero*, malformation of structures other than the human central nervous system seems to be uncommon in the human species, although common, at least at high doses, in experimental animals. However, it has been found that the development of the human central nervous system may be affected adversely by low radiation doses, with high, though variable susceptibility through a lengthy period of time during fetal development (Otake and Schull, 1984).

The UNSCEAR (1986) has attempted to develop quantitative risk estimates for several effects of prenatal irradiation on the embryo-fetus in relation to the period of pregnancy over which the specific risk applies, using several qualifying assumptions. Mental retardation related to prenatal irradiation is not seen in children irradiated up to eight weeks after fertilization nor after 25 weeks after fertilization. For irradiation at 8 to 15 weeks after fertilization the probability of induction of mental retardation by prenatal irradiation appears to fit a linear relationship with dose, with no observed dose threshold in the available data. The probability of induction of mental retardation in this period has a mean value of approximately 0.4 Sv^{-1} (4×10^{-3} rem^{-1}) over a fetal dose equivalent range of 0.01 to more than 2 Sv (1 to more than 200 rem). Between the 16th and 25th week after conception, the dose response curve may be more complex but the risk of mental retardation is less and can be taken to be about 0.1 Sv^{-1} (1×10^{-3} rem^{-1}). According to scores from objective tests of intelligence or mental capability, the severity of the radiation induced mental deficit following irradiation in these susceptible periods during gestation appears to increase with dose from slight to marked.

The UNSCEAR's (1986) value for the total risk for the period of gestation for several effects of irradiation of the embryo-fetus (mortality, induction of malformations, mental retardation, solid tumors and leukemia) is about 0.2 Sv^{-1} (2×10^{-3} rem^{-1}) as an upper limit. UNSCEAR indicates that for the smaller doses likely to be encountered in practice, the risk is thought to be relatively small in relation to the natural incidence or probability, which is assumed to be of the order of 6×10^{-2} for anomalies of all kinds that may seriously affect health and viability of newborn children.

The NCRP recommends a total dose equivalent limit (excluding medical exposure) of 5 mSv (0.5 rem) for the embryo-fetus. Once a pregnancy becomes known, exposure of the embryo-fetus shall be no greater than 0.5 mSv (0.05 rem) in any month (excluding medical exposure).

This recommendation is intended to limit exposure to a fetus of an occupationally exposed mother. However, the special sensitivity of the fetus should be considered in all situations, including those involving medical exposure of the expectant mother. (See also ICRP, 1984a, 1986.)

The total dose equivalent limit, for the whole period of pregnancy, is important for the limitation of the risk of cancer induction and of developmental anomalies, while the monthly limit is important to ensure that exposures of the embryo-fetus during particularly critical periods of organogenesis and development are adequately restricted. (The dose equivalent and the effective dose equivalent for the fetus are taken to be the same in this Report.)

For purposes of radiation protection, the NCRP regards the embryo-fetus as a separate entity, distinct from the woman bearing it. Thus, the embryo-fetus carried by a woman who is a radiation worker is not regarded as subject to the occupational limits, but rather to lower limits that are necessary for its adequate protection. It is recognized that in the occupational case, the recommended limits for the embryo-fetus may place special restrictions on women who are known to be pregnant. This may pose problems concerning matters such as disclosure of potential and actual pregnancy, risk-sharing among workers, associated economic problems, equal employment considerations, and perhaps other legal matters. Some of these problems are discussed in somewhat greater length in NCRP Report No. 53 (NCRP, 1977a).

It is implied in the NCRP's recommendations that women who are capable of becoming pregnant should not be exposed to radiation environments in which the uterus could receive a dose equivalent in excess of the total fetal limit, 5 mSv (0.5 rem), before a pregnancy becomes known. *Therefore, special attention is required to ensure that, if exposures are received occupationally, they are distributed uniformly with time so that the embryo-fetus does not receive more than its limit before pregnancy is known.*

It is the intention of the NCRP that the dose equivalent to the embryo-fetus, rather than the maternal dose equivalent, be the basis for implementation of the recommended limits for the embryo-fetus. When the exposure is restricted to radiation of low penetrating power, the recommended dose equivalent limits for the embryo-fetus may be satisfied if the annual maternal dose equivalent, as measured by external dosimetry, is within the general occupational dose equivalent limit, if acquired at a fairly uniform rate. If the radiation is of high penetrating power, however, the difference between the maternal dose equivalent, and the dose equivalent to the embryo-fetus will be much smaller than in the case of radiation of low penetrating power. In this case, it is especially important to estimate the dose equivalent to the embryo-fetus (NCRP, 1977b) in order to assure that the recommended embryo-fetus limit has not been exceeded.

Internally deposited radionuclides pose special problems for protection of the embryo-fetus, and particular care should be taken with them because some remain in the body for long periods of time and their transfer and the doses delivered to fetal organs are not well known in all cases. Therefore, it is important to prevent intakes of radionuclides by fertile women that could give to a present or future embryo-fetus a dose equivalent that would exceed the recommended limit. The NCRP has a Task Group addressing the problem of radionuclides and the embryo-fetus. Also, the NCRP has a Committee concerned with the effects of radiation on the embryo-fetus which will continue to examine information relevant to the protection of the embryo-fetus.

12. Exposure in Excess of the Limits: Occupational

The annual limits are intended to control the maximum risk to an individual in any year. Slightly exceeding the annual dose equivalent limit has little biological significance for the individual since the lifetime risk will be only slightly perturbed and can be readily offset by reduced exposure in future years. The importance of such occurrences is that they call attention to what may be an inadequate system of radiation control. This would suggest that the decision on permitting individuals who have exceeded dose limits to return to radiation worker status depends more upon the improvement and corrections in control of radiation exposure in the work place than on in-depth analysis of the worker's health status.

Annual Limits on Intake (ALIs) are intended to control radionuclide exposures. Occasional intakes slightly greater than the ALIs will have little effect on the long term health of an individual. However, in some cases it may be desirable to estimate the resulting annual effective dose equivalent as well as the committed effective dose equivalent. Since the ALIs were calculated for the ICRP Reference Man, estimates of dose equivalent for individuals may be more accurate for assessment purposes if derived from information specific to the nature of the intake, such as the physical and chemical properties of the radionuclide, *in vivo* counting or bioassay data and the age and other physiological parameters of the exposed individual. The principal significance of most radionuclide intakes that exceed the ALIs is their indication of the need for improvements in operating procedures.

13. Dose Equivalent Limits for Unusual Occupational Situations

In the development of the recommendations pertaining to radiation workers, the average risk in safe industries has been used as the basis. The Council believes that most radiation occupations and industries can adhere to this risk level without undue difficulty, and many can operate at a much lower risk level due to their relatively low occupational exposure. There may exist, however, unusual occupational situations in which the worker population cannot carry out required functions under the recommended annual limits, and indeed, risks other than radiation may be much larger than normal. Such special circumstances may require special limits and a different basis of comparison. The radiation risks in these circumstances could be compared with those from the less safe industries such as public utilities and transportation, construction, agriculture and mining. In these situations, the main concern also would be for the accumulated exposure of individuals over their working lifetime and thus a career limit would be of special importance. Space flight activities, especially for long missions, is an example of such circumstances.

14. Reference Levels: Occupational

As stated earlier, neither annual effective dose equivalent limits nor ALIs are generally suitable in themselves for design and control because ALARA is also part of the radiation protection system. Thus, there is a need to develop reference levels which reflect ALARA considerations. For example, if it can be demonstrated that operations at a given facility can be carried out without exceeding a certain weekly level below that corresponding to the annual limits, the management might well establish a periodic reference level based on this information. This would serve the objective of keeping employees cognizant of the obligation to maintain their exposure as low as reasonably achievable and drawing attention to the constant need for good radiation protection practices. If an exposure exceeded this value, a predetermined course of action could be taken, such as investigation, retraining, maintenance, or repair. (It may be appropriate to specify different reference levels for each of these actions.) The establishment of reference levels is most properly done by appropriate management at each facility because the data required will be site-specific.

15. Planned Special Exposure: Occupational

In order to allow certain essential tasks to be performed, it may be necessary, *on rare occasions*, to permit a few workers to receive an annual effective dose equivalent in excess of 50 mSv (5 rem). *The NCRP recommends that if exposure under these conditions is essential, these guidelines be followed:*

(1) No worker should receive an effective dose equivalent[10] of more than 100 mSv (10 rem) in a single planned event,

(2) Nor should the effective dose equivalent received in such special planned exposures exceed 100 mSv (10 rem) over the working lifetime,

(3) Such planned special exposures should be authorized in writing by senior management prior to the exposure,

(4) Older workers with low lifetime effective dose equivalents should be selected whenever possible,

(5) Exposures resulting from planned special exposures shall be included in the lifetime record of exposure for each worker but separately identified.

[10] The effective dose equivalent specified here must be the sum of external and internal effective dose equivalent, if both exist.

16. Guidance for Emergency Occupational Exposures

Only actions involving life saving, justify acute exposures in excess of 100 mSv (10 rem). The use of volunteers for exposures during emergency actions is desirable. Older workers with low lifetime accumulated effective dose equivalents should be chosen from among the volunteers whenever possible. Exposure during emergency actions which do not involve life saving should be controlled by the occupational exposure limits, including the planned special exposure recommendations.

When the exposure may approach or exceed 1 Gy (100 rad) of low-LET radiation (or an equivalent high-LET exposure) to a large portion of the body, in a short time, the worker needs to understand not only the potential for acute effects but he or she should also have an appreciation of the substantial increase in his or her lifetime risk of cancer. If the possibility of internal exposures also exist, these must be taken into account.

17. Nonoccupational Limits: Exposure of Members of the Public

The development of recommendations on limitation of exposure of the public is more difficult than in the occupational case because comparisons such as those used in that case cannot be as explicit. Thus, a certain degree of arbitrariness is inevitable. Nevertheless, there are pertinent considerations upon which to base reasonable judgments. First, the risks associated with a given dose are as discussed in Section 7. Second, the manifold mortality risks faced by members of the public vary greatly, but are commonly in the range of about 10^{-4} to 10^{-6} per year, and sometimes higher (see, for example, ICRP, 1977a; Pochin, 1981; Wilson and Crouch, 1982; Sinclair 1985a). Not only do these risks exist, but they seem, depending somewhat on their nature, to be accepted. Third, everyone is exposed to natural background radiation, and that, annually, is commonly about 1 mSv (100 mrem) (excluding radon) or a risk of mortality of about 10^{-5} annually, or approximately 10^{-3} lifetime. The annual effective dose equivalent from background varies in the United States from about 0.65 mSv (65 mrem) on the Atlantic Seaboard to 1.25 mSv (125 mrem) in Denver. The average annual effective dose equivalent due to radon is about 2 mSv (200 mrem) and variations in it are much greater (NCRP, 1984a) than the average value of natural background from other sources.

It is necessary to distinguish between exposures resulting from man-made sources and those resulting from natural background. Potential means of control of these two types of sources differ significantly. In many cases natural sources are difficult to control. Indeed, it is only in recent times that it has been recognized that some natural sources need to be controlled. The recommendations set out below treat the two types of sources separately.

On the basis of the above considerations, the following levels are recommended for man-made sources other than medical.[11] *For contin-*

[11] Medical exposure refers to that exposure received by individual patients as a result of medical procedures performed on them.

uous (or frequent) exposure, it is recommended that the annual effective dose equivalent not exceed 1 mSv (0.1 rem). This recommendation is designed to limit the exposure of members of the public to reasonable levels of risk comparable with risks from other common sources, *i.e.*, about 10^{-5} annually. Furthermore, *a maximum annual effective dose equivalent limit of 5 mSv (0.5 rem) is recommended to provide for infrequent annual exposures (NCRP, 1984a).* This recommendation is made because annual exposures in excess of this limit, to a small group of people, need not be regarded as especially hazardous, provided they do not occur often to the same groups.

Both the 1 mSv (0.1 rem) and the 5 mSv (0.5 rem) annual effective dose equivalent limits for members of the public will keep the annual dose equivalent to those organs and tissues that are considered in the effective dose equivalent system below levels of concern for nonstochastic effects. However, because some organs and tissues are not routinely included in the calculation of effective dose equivalent, *i.e.*, extremities, skin and lens of eye (see Section 5), an annual dose equivalent limit of 50 mSv (5 rem) is recommended for these organs or tissues.

The ALIs for radionuclides published by the ICRP (1979) are for workers and are based on committed effective dose equivalents. For adult members of the public exposed to radionuclides in the same physical and chemical form as those tabulated for workers (ICRP, 1979), it is appropriate to divide the ALI values by 10 for meeting the stochastic dose limitations for infrequent exposures and by 50 for continuous exposures. For other situations involving exposure of members of the public, individual factors such as chemical and physical form of the radionuclides, body size and metabolism should be taken into account. Some guidance is offered in ICRP Publication 39 (ICRP, 1984a).

Notes:

(1) The recommendations apply to exposures from man-made sources and exclude medical and natural background exposures. Medical exposures are excluded from this limitation because they confer personal benefit to the exposed individual and are extremely variable from one person to another. Guidelines exist for the exposure of the public to medical procedures and these should be applied (EPA, 1978; FDA, 1985a, 1985b). Natural background exposure is excluded also. Natural background radiation is ubiquitous and variations in natural background levels (excluding radon) are small. The first recommendation effectively permits the exposure of an individual from man-made sources to equal that from average natural background (exclud-

ing radon). Natural background exposure is, however, included in recommendations specifying remedial action levels (see Section 18).

(2) Exposures controlled by these annual limits, 1 mSv (0.1 rem) continuous and 5 mSv (0.5 rem) infrequent, are subject to the application of ALARA considerations in the same way as for occupational exposures.

(3) When exposures are from internal and external sources, the contributions should be summed in a manner similar to that given for occupational exposures in Section 10, so that the total exposure does not exceed the respective limits.

In some special circumstances, including emergencies, small segments of the public may be exposed to radiation in excess of the recommended annual limit for infrequent exposure, *i.e.*, 5 mSv (0.5 rem). In this event, the NCRP recommends that the general guidelines relating to planned special exposures for workers be employed. In some very special circumstances, the guidelines for emergencies, given in Section 16, may also be applicable.

18. Remedial Action Levels for Members of the Public

If the recommendations of the previous Section are observed, man-made radiation sources will not expose members of the public to annual effective dose equivalents greater than 1 mSv (0.1 rem) continuously, or 5 mSv (0.5 rem) infrequently. Exposures should always be less than the limits and, indeed, on the average, utilizing the principles of ALARA, they should be much less.

However, natural background is excluded from those limits and there are circumstances in which natural background itself, or more especially, natural radiation sources enhanced locally by man's operations for selected purposes, can give rise (sometimes quite inadvertently) to annual exposures above the level of 1 mSv (0.1 rem).

It then becomes necessary to consider at what exposure level remedial action, which may be possible only at substantial societal cost, should be undertaken. Remedial action levels involve a balance of risk and many other socioeconomic factors. In general, the aim of setting a remedial action level is to reduce the greatest risks from a given type of radiation source. It is clear that once a remedial action level is established for given circumstances, action is mandatory when a level above it is found. Actions to reduce exposure should not be limited by or to the remedial action level and, following the ALARA principle, levels substantially below the remedial action level may be obtainable and appropriate.

For external sources, the NCRP considers that the risks to the public from exposure to all sources except medical, should not exceed about five times the total of other risks faced by members of the public. Thus, an annual remedial action level is specified at an effective dose equivalent of 5 mSv (0.5 rem) for *all external sources* other than medical. External sources are specified because internal exposures from radionuclides other than radon are rarely limiting in present circumstances (NCRP, 1984a).

The recommended remedial action level, 5 mSv (0.5 rem), is 10 times greater than the average annual effective dose equivalent due to external exposure from natural background 0.5 mSv (0.05 rem). It is also comparable with the annual effective dose equivalent received by many radiation workers.

The NCRP has given special attention to the problems occasioned by exposure to indoor radon (NCRP, 1984a, 1984b) and notes that *this is potentially the most important public radiation exposure problem that currently exists.* As a result, a remedial action level of 0.007 Jhm^{-3} y^{-1} (2 WLM y^{-1}) was recommended in NCRP Report No. 77 (NCRP, 1984a). Elements of feasibility enter the considerations here since it is evident from NCRP Report No. 77 (NCRP, 1984a) that in a substantial number of homes the radon levels are estimated to exceed the average by amounts up to 5 or 10 times or more. It is certainly desirable that such levels be reduced and the risks associated with them decreased. A remedial action level must, therefore, be chosen for which the societal impacts are not excessive, but the greatest risks are avoided. The NCRP recognizes that an annual inhalation level for radon that corresponds to 5 mSv (0.5 rem) effective dose equivalent would be about 0.00175 Jhm^{-3} (0.5 WLM), see ICRP Publication 32 (ICRP, 1981). However, this is only two and one-half times the present estimated average annual indoor radon background exposure of 0.0007 Jhm^{-3} (0.2 WLM) and imposition of a remedial action level at this value could involve a very large number of homes. Therefore, the NCRP proposed a remedial action level which was based on excess lifetime risk being no more than 10 times the present average annual background level, or 0.007 Jhm^{-3} y^{-1} (2 WLM y^{-1}) (NCRP, 1984a). The annual risk of lung cancer associated with this level is 4×10^{-4}, and NCRP considers risks of this magnitude undesirable. However, it is anticipated that remedial actions, once taken, will, together with ALARA, establish new annual radon exposures in a given home much below 0.007 Jhm^{-3} (2 WLM).

It is also anticipated, over time, and assuming that the problem of indoor radon is addressed by taking the worst situations first, that radon levels in existing homes will be reduced. Furthermore, the Council believes that for new homes, suitable constraints should be developed so that they will have radon levels below those of many present structures.

For the present, it is recommended that remedial action be undertaken:

(1) *When the average annual effective dose equivalent from external exposure*[12] *(excluding medical, but including naturally occurring sources) continuously exceeds 5 mSv (0.5 rem).*

(2) *When the total exposure to radon and its decay products for an individual exceeds an annual average of 0.007 Jhm^{-3} (2 WLM).*

[12] In the unlikely event, that internal exposure other than radon could make a significant contribution, it should be included in the assessment of exposure.

19. Control of Sources: Exposure of Members of the Public

The control of individual sources which can result in exposures of members of the public to low levels of radiation is not usually the subject of NCRP recommendations, provided the resulting exposures do not exceed the recommended limits. However, in response to a specific need, the NCRP addressed the subject of control of multiple sources in the case of airborne radionuclides (NCRP, 1984c). The NCRP reaffirms that recommendation as follows; whenever the potential exists for an individual (member of the public) to exceed 25 percent of the annual effective dose equivalent limit from irradiation attributable to any single site, the site operator should be required to assure that the annual exposure of the maximally exposed individual from all sources would not exceed 1 mSv (100 mrem) on a continuous basis. If each site operator conforms to this requirement, the limits of exposure to the public should not be exceeded.

20. Negligible Individual Risk Level (NIRL)

A negligible individual risk level (NIRL) is defined here as a level of average annual excess risk of fatal health effects attributable to irradiation, below which further effort to reduce radiation exposure to the individual is unwarranted. The NIRL is regarded as *trivial* compared to the risk of fatality associated with ordinary, normal societal activities and can, therefore, *be dismissed from consideration.*

Whether or not annual levels of irradiation for individuals at or below the NIRL are measurable with certainty, they can be estimated or calculated. Such levels, at or below the NIRL, are not required to be considered for purposes of radiation protection. The NIRL is regarded as a threshold below which the control of radiation sources and limitation of exposures, *i.e.*, efforts to reduce risk further, would be deliberately and specifically curtailed. In view of the relative numbers of people in the general population, as compared with the numbers of radiation workers, the utilization of the NIRL is especially important in regard to environmental issues involving exposure of populations.

It should be emphasized that the NIRL is not a radiation exposure limit and that exposure to small effective dose equivalent increments above those corresponding to the NIRL are of little significance. *The NIRL should not be confused with an acceptable risk level, a level of significance or a limit.* The NCRP has recommended an annual effective dose equivalent limit for continuous exposure of members of the public at 1 mSv (0.1 rem) on the basis of the comparability of the associated risk with societal risks that appear to be accepted. Levels of exposure less than this limit become progressively less significant with respect to risk as the annual effective dose equivalent decreases, finally becoming completely insignificant at a level corresponding to the NIRL.

The NIRL is not the goal of ALARA. However, it does provide a lower limit for application of the ALARA process. In principle, ALARA requires efforts to keep radiation exposure as far below that permitted by the limits as is reasonably achievable, economic and social factors being taken into account. The concept of ALARA does not justify attempts to reduce detriment to levels so low or trivial as to be inappropriate in relation to reasonable priorities for expenditure of health protection resources. The relationships between the NIRL, ALARA, and maximum permissible levels are emphasized here because of the tendency for established numerical guidelines to be interpreted inappropriately as upper limits.

In the radiation protection field, the need for a reasonably negligible risk level to avoid excessive control actions and expenditures to reduce individual risk has long been recognized. In effect, regulatory practices have involved low exposure cut-off levels for various hazardous agents or situations, apparently on the basis of the triviality of the risks as compared with other natural or man-made risks. Decisions based on the concept of reasonably negligible risk are made frequently in many societal and individual activities.

The NIRL should be assuredly low, in view of the likely variation in susceptibility to radiation effects among individuals in heterogeneous human populations and uncertainties about risk levels. The NIRL should also be regarded as adjustable on the basis of new and better information which may bear on its numerical value.

In deriving the recommended numerical value for the NIRL in this Report, several criteria relevant to smallness or triviality of risk were considered which, taken together, offer degrees of reasonableness and perspective that tend to minimize subjective aspects of judgment. Smallness of risk was considered in relation to: (1) magnitude of dose; (2) difficulty in detection and measurement of dose and health effect; (3) natural risk for the same health effects; (4) estimated risk for the mean and variance of natural background radiation exposure levels; (5) risks to which people are accustomed; and (6) perception of, and behavioral response to, risk levels. The limiting of radiation risk among radiation workers, to levels of risk in the occupations generally regarded as safe, is a valuable objective approach. This kind of approach, at a lower level of comparable risk, contributes to the establishment of a value for the NIRL.

As indicated in Section 17, the recommended annual effective dose equivalent limit of 1 mSv (0.1 rem) for continuous (or frequent) exposure for members of the public is designed to limit their exposure to reasonable levels of risk comparable with risks from other common sources. The risk from such exposure at that limit would be approximately 10^{-5} annually or 10^{-3} lifetime. Correspondingly, an average annual risk of 10^{-6} would result from an annual exposure of 0.1 mSv (0.01 rem), and an annual risk of 10^{-7} from an annual exposure of 0.010 mSv (0.001 rem). The accumulation of risk at 10^{-7} annually corresponds with a lifetime (70 y) risk of about 0.7×10^{-5}, a value that is about one one-hundredth of the risk corresponding to the limit for continuous exposure of members of the general public. It is again emphasized that the very small risks estimated in this section cannot be demonstrated but result from the assumption of the linear, no threshold relationship between radiation dose equivalent and stochastic effects.

On the basis of the considerations of the collective criteria mentioned above, an estimated and assumed lifetime risk of 10^{-5} for fatal health effects of radiation exposure, or annual risk commitment increments of 10^{-7}, are taken to be negligible. For continuous or repeated exposure, this level of risk corresponds with annual dose equivalent increments of about 0.01 mSv (0.001 rem). This dose equivalent rate is about one percent of the recommended limit for continuous irradiation of members of the general public, and about one percent of the average natural background radiation exposure excluding radon.

The NCRP recommends recognition of a level of average excess fatal health risks from radiation exposure from any individual source or practice below which further effort to reduce individual exposure is unwarranted. *That level of risk is defined as the Negligible Individual Risk Level (NIRL) and is given an annual value of* 10^{-7}, *the annual risk that corresponds to an annual effective dose equivalent of 0.01 mSv (0.001 rem). It is further recommended that assessments of increments of collective annual effective dose equivalents from any particular individual source or practice should exclude those individuals whose annual effective dose equivalent from such a source is 0.01 mSv (0.001 rem) or less.*

It is inherent in the foregoing recommendation that the annual exposure resulting from each source or practice which does not exceed 0.01 mSv (0.001 rem) is dismissible from consideration. As individuals may be exposed to more than one radiation source or practice, it is possible that the total of the annual individual radiation exposure dismissible on the basis of the NIRL principle could exceed that associated with the annual NIRL *per se*. In practice, however, it is unlikely that an individual would be exposed to radiation from as many as 10 sources in any year. Accordingly, assuming exposure of an individual to a maximum of 10 radiation sources or practices in a year, the use of the NIRL could result in the annual dismissal of up to 0.1 mSv (0.01 rem) for an individual.

21. Trainees Under 18 Years of Age

For educational and training purposes, it may be necessary and desirable to accept occasional exposure of persons under the age of 18 years. *It is recommended that exposures for these purposes be permitted only under conditions presenting high assurance of maintaining the resulting annual effective dose equivalent to less than 1 mSv (0.1 rem).* This is considered to be a part of the annual limit of 5 mSv (0.5 rem) given in Section 17 for infrequent exposure for members of the public and not supplemental to it. Intentional exposure of trainees should be avoided.

It is recognized that a productive part of the training experience may be better conducted in an industrial or hospital situation, which might constitute part-time work experience, supervised in some manner by an educational institution. The NCRP recommends that such work experience be governed by the radiation protection practices recommended for educational institutions in NCRP Report No. 32 (NCRP, 1966), with, however, the revision in limits recommended here.

22. Summary of Recommendations

A number of important modifications of the Council's earlier recommendations on basic radiation protection criteria (NCRP, 1971a) have been presented. These modifications are as follows.

For occupational exposures: first, the NCRP recommends discontinuance of the age proration formula of 5 (N-18) rem where N is age in years. Second, the NCRP continues the use of the annual limit of 50 mSv (5 rem) but emphasizes its boundary nature. Third, it is suggested, as guidance for radiation protection programs, that cumulative exposure should not exceed the age of the individual in years × 10 mSv (years × 1 rem). Fourth, in the exposure of pregnant women under occupational conditions, it is recommended that the limit for the fetus [5 mSv (0.5 rem)] not be received at a rate greater than 0.5 mSv (0.05 rem) per month. Fifth, it is recommended explicitly that all limits include the sum of external and internal exposures.

For public exposures, to man-made sources, *i.e.*, other than medical and natural background, NCRP reaffirms the annual limits given in NCRP Report No. 77 (NCRP, 1984a) of 1 mSv (0.1 rem) for continuous exposures and 5 mSv (0.5 rem) for infrequent exposures. Also reaffirmed are remedial action levels for the public of 5 mSv (0.5 rem) annual average for external exposure (from all sources except medical) and an annual average of 0.007 Jhm^{-3} (2 WLM) for total exposure to radon and its decay products.

In this Report, the NCRP defines an annual Negligible Individual Risk Level (NIRL) which establishes a boundary below which the risk is dismissed from consideration: *i.e.*, at a risk of 10^{-7}, corresponding to a dose equivalent of 0.01 mSv (0.001 rem).

All of the current recommendations on limits are summarized in Table 22.1.

Another important change is the assignment of new values to the effective quality factors for some radiations (*e.g.*, 20 for neutrons and alpha particles of unknown energy) see Table 4.1.

It needs to be recognized that the recommendations given here may require some modification as additional information becomes available. However, the NCRP has attempted to make the judgements reflected in these recommendations as forward looking as possible.

TABLE 22.1—*Summary of recommendations*[a,b]

A. Occupational exposures (annual)[c]		
1. Effective dose equivalent limit (stochastic effects)	50 mSv	(5 rem)
2. Dose equivalent limits for tissues and organs (nonstochastic effects)		
a. Lens of eye	150 mSv	(15 rem)
b. All others (*e.g.*, red bone marrow, breast, lung, gonads, skin and extremities)	500 mSv	(50 rem)
3. Guidance: Cumulative exposure	10 mSv × age	(1 rem × age in years)
B. Planned special occupational exposure, effective dose equivalent limit[c]	see Section 15	
C. Guidance for emergency occupational exposure[c]	see Section 16	
D. Public exposures (annual)		
1. Effective dose equivalent limit, continuous or frequent exposure[c]	1 mSv	(0.1 rem)
2. Effective dose equivalent limit, infrequent exposure[c]	5 mSv	(0.5 rem)
3. Remedial action recommended when:		
a. Effective dose equivalent[d]	>5 mSv	(>0.5 rem)
b. Exposure to radon and its decay products	>0.007 Jhm^{-3}	(>2 WLM)
4. Dose equivalent limits for lens of eye, skin and extremities[c]	50 mSv	(5 rem)
E. Education and training exposures (annual)[c]		
1. Effective dose equivalent limit	1 mSv	(0.1 rem)
2. Dose equivalent limit for lens of eye, skin and extremities	50 mSv	(5 rem)
F. Embryo-fetus exposures[c]		
1. Total dose equivalent limit	5 mSv	(0.5 rem)
2. Dose equivalent limit in a month	0.5 mSv	(0.05 rem)
G. Negligible Individual Risk Level (annual)[c]		
Effective dose equivalent per source or practice	0.01 mSv	(0.001 rem)

[a] Excluding medical exposures.
[b] See Table 4.1 for recommendations on $\bar{Q}$.
[c] Sum of external and internal exposures.
[d] Including background but excluding internal exposures.

Important contributions to future modifications will be made by studies concerned with relative biological effectiveness at low doses, the shape of dose response curves at low doses, their dependence on dose rate and considerations of the many agents that can modify the carcinogenic response. Additional information on human risks of radiation exposure and the projection models to be used with them and on weighting factors for use in radiation protection can be expected to become available in the future. The NCRP will follow developments in these areas closely and use them to form the basis of new or revised recommendations when necessary.

Glossary

absorbed dose: The quotient of $d\overline{E}$ by dm where $d\overline{E}$ is the mean energy imparted by ionizing radiation to the matter in a volume element and dm is the mass of the matter in that volume element, *i.e.*, the absorbed dose, $D = d\overline{E}/dm$. The special name for the unit of absorbed dose ($J\ kg^{-1}$) is the gray (Gy). The earlier unit of absorbed dose was the rad which is equal to 0.01 Gy.

ALARA: *A*s *L*ow *A*s *R*easonably *A*chievable, economic and social factors being taken into account.

annual limit on intake (ALI): The activity of a radionuclide that, taken into the body during a year, would provide a committed effective dose equivalent to a person, represented by Reference Man, equal to the annual occupational effective dose equivalent limit ($H_{E,L}$) or, in some cases, the organ dose equivalent limit (nonstochastic effect). The ALI is normally expressed in becquerels (Bq) or curies.

becquerel (Bq): The special name for the unit of activity in the SI. $1\ Bq = 1\ s^{-1}$.

committed dose equivalent (H_{50}): The dose equivalent accumulated in the 50 years after intake of a radionuclide.

derived air concentration (DAC): The ALI of a radionuclide divided by the volume of air inhaled by Reference Man in a working year (*i.e.*, $2.4 \times 10^3\ m^3$). The unit of DAC is $Bq\ m^{-3}$.

dose equivalent (H): A quantity used for radiation protection purposes that expresses on a common scale for all radiations, the irradiation incurred by exposed persons. It is defined as the product of the absorbed dose (D) and the quality factor (Q). The name for the unit of dose equivalent ($J\ kg^{-1}$) is the sievert (Sv).

dose equivalent to whole body (H_{wb}): See whole body dose equivalent.

effective dose equivalent (H_E): The sum over specified tissues of the products of the dose equivalent in a tissue (T) and the weighting factor for that tissue, (w_T), *i.e.*, $H_E = \sum w_T H_T = H_{wb}$.

gray (Gy): The special name for the SI unit of absorbed dose, kerma and specific energy imparted. $1\ Gy = 1\ J\ kg^{-1} = 100$ rad.

negligible individual risk level (NIRL): A level of risk that can be dismissed. Namely, an annual risk of 10^{-7}. This risk is that associated with an annual effective dose equivalent of 0.01 mSv (0.001 rem).

nonstochastic effects: Effects for which the severity of the effect in affected individuals varies with the dose, and for which a threshold usually exists.

optimization: This has the same meaning as ALARA.

organ weighting factor (w_T): A factor that indicates the ratio of the risk of stochastic effects attributable to irradiation of a given organ or tissue (T) to the total risk when the whole body is uniformly irradiated.

quality factor (Q): A factor used for radiation protection purposes that accounts for differences in biological effectiveness between different radiations. It is the ratio of the slope of the curve of risk versus dose for a given radiation to that of a reference radiation in the range of dose where the curves are assumed to be linear (see Section 4).

rad: A special unit for absorbed dose, kerma, and specific energy imparted. One rad is 0.01 joules absorbed per kilogram of any material. (Also defined as 100 ergs per gram.) Being replaced by the gray. 1 rad equals 0.01 gray.

reference level: The predetermined value of a quantity, below a limit, which triggers a specified course of action when the value, usually a dose level, is exceeded or is expected to be exceeded.

rem: A special unit for dose equivalent. Being replaced by the sievert, 1 rem = 0.01 Sv.

secondary limit: A limit, derived from a primary limit using conservative assumptions, that assures adherence to the primary limit by methods easier to implement than those required for the primary limit.

sievert (Sv): The special name for the SI unit of dose equivalent. 1 Sv = 1 J kg^{-1} = 100 rem.

stochastic effects: Effects, the probability of which, rather than their severity, is a function of radiation dose without threshold. (More generally, stochastic means random in nature.)

teratogenic effects: Effects occurring in offspring as a result of insults sustained in-utero.

whole body dose equivalent (H_{wb}): The dose equivalent associated with the uniform irradiation of the whole body. $H_{wb} = H_E$.

working level (WL): That amount of potential alpha energy in a cubic meter of air that will result in the emission of 2.08×10^{-5} joules of energy.

working level month (WLM): A cumulative exposure, equivalent to exposure to one working level for a working month (170 hours), *i.e.*, 2.08×10^{-5} $Jm^{-3} \times 170$ h = 0.0035 Jhm^{-3}.

References

Broerse, J. J., Hennen, L. A. and van Zwieten, J. M. (1985). "Radiation carcinogenesis in experimental animals and its implications for radiation protection," Int'l J. Radiat. Biol. **48**, 167.

Dennis, J. A. and Dunster, H. J. (1985). "Radiation quality and radiation protection: implications of changes in quality factors," Radiat. Prot. Dosimetry **13**, 327.

DOE (1984). Department of Energy. *U.S. Department of Energy Injury and Property Damage Summary, January-March, 1984*, Report by E.G. & G. Idaho, Inc. to U.S. DOE Assistant Secretary for Environment, Safety and Health, reported to DOE annually under contract DE-AC07-761DO1570 (National Technical Information Service, Springfield, Virginia).

EPA (1978). Environmental Protection Agency. *Radiation Protection Guidance to Federal Agencies for Diagnostic X Rays*, page 4377, Federal Register **43**, No. 22.

EPA (1984). Environmental Protection Agency. *Occupational Exposure to Ionizing Radiation in the United States, A Comprehensive Review for the Year 1980 and a Summary of Trends for the Years 1960–1985*, EPA 520/1-84-005 (National Technical Information Service, Springfield, Virginia).

FDA (1985a). Food and Drug Administration. *Evaluation of Radiation Exposure from Diagnostic Radiology Examinations, General Recommendations*, HHS Publication (FDA) 85-8246 (National Technical Information Service, Springfield, Virginia).

FDA (1985b). Food and Drug Administration. *Recommendations for Evaluation of Radiation Exposure from Diagnostic Radiology Examinations*, R. L. Burkhart, R. E. Gross, R. G. Jans, J. L. McCrohen, Jr., M. Rosenstein and F. A. Reuter, Eds., HHS Publication (FDA) 85-8247 (National Technical Information Service, Springfield, Virginia).

FRC (1960). Federal Radiation Council. *Radiation Protection Standards*, FRC Report No. 1 (National Technical Information Service, Springfield, Virginia).

ICRP (1969). International Commission on Radiological Protection. *Radiosensitivity and Spatial Distribution of Dose*, ICRP Publication 14 (Pergamon Press, New York).

ICRP (1973). International Commission on Radiological Protection. *Data for Protection Against Ionizing Radiation from External Sources*, ICRP Publication 21 (Pergamon Press, New York).

ICRP (1975). International Commission on Radiological Protection. *Reference Man: Anatomical, Physiological and Metabolic Characteristics*, ICRP Publication 23 (Pergamon Press, New York).

ICRP (1977a). International Commission on Radiological Protection. *Recommendations of the International Commission on Radiological Protection*, ICRP Publication 26 (Pergamon Press, New York).

ICRP (1977b). International Commission on Radiological Protection. *Problems Involved in Developing an Index of Harm*, ICRP Publication 27 (Pergamon Press, New York).

ICRP (1978). International Commission on Radiological Protection. *Statement from the 1978 Stockholm Meeting of the ICRP*, in ICRP Publication 28 (Pergamon Press, New York).

ICRP (1979). International Commission on Radiological Protection. *Limits for Intakes of Radionuclides by Workers*, ICRP Publication 30 (Pergamon Press, New York).

ICRP (1980). International Commission on Radiological Protection. *Statement and Recommendations of the 1980 Brighton Meeting of the ICRP*, in ICRP Publication 30 (Pergamon Press, New York).

ICRP (1981). International Commission on Radiological Protection. *Limits for Inhalation of Radon Daughters by Workers*, ICRP Publication 32 (Pergamon Press, New York).

ICRP (1983). International Commission on Radiological Protection. *Cost-Benefit Analysis in the Optimization of Radiation Protection*, ICRP Publication 37 (Pergamon Press, New York).

ICRP (1984a). International Commission on Radiological Protection. *Statement from the 1983 Washington Meeting of the ICRP*, in ICRP Publication 39 (Pergamon Press, New York).

ICRP (1984b). International Commission on Radiological Protection. *Nonstochastic Effects of Ionizing Radiation*, ICRP Publication 41 (Pergamon Press, New York).

ICRP (1985a). International Commission on Radiological Protection. *Quantitative Bases for Developing a Unified Index of Harm*, ICRP Publication 45 (Pergamon Press, New York).

ICRP (1985b). International Commission on Radiological Protection. *Statement From the 1985 Paris Meeting of the ICRP*, Health Phys. **48**, 828.

ICRP (1986). International Commission on Radiological Protection. *Developmental Effects of Irradiation on the Brain of the Embryo and Fetus*, ICRP Publication 49 (Pergamon Press, New York).

ICRP-ICRU (1963). International Commission on Radiological Protection and the International Commission on Radiation Units and Measurements. "Report of the RBE Committee to the International Commission on Radiological Protection and the International Commission on Radiation Units and Measurements," Health Phys. **9**, 357.

ICRU (1970). International Commission on Radiation Units and Measurements. *Linear Energy Transfer*, ICRU Report No. 16 (International Commission on Radiation Units and Measurements, Bethesda, Maryland).

ICRU (1986). International Commission on Radiation Units and Measurements. *The Quality Factor in Radiation Protection*, Report of a Joint Task Group of the ICRP and the ICRU to the ICRP and the ICRU, ICRU Report 40 (International Commission on Radiation Units and Measurements, Bethesda, Maryland).

KATO, H. AND SCHULL, W. J. (1982). "Studies of the mortality of A-bomb survivors, 7. Mortality, 1950–1978: Part I. Cancer mortality," Radiat. Res. **90**, 395.

Lewis, E. B. (1957). "Leukemia and ionizing radiation," Science **125**, 965.

NAS/NRC (1972). National Academy of Sciences/National Research Council. *The Effects on Populations of Exposure to Low Levels of Ionizing Radiation,*

Report of the Advisory Committee on the Biological Effects of Ionizing Radiation (National Academy Press, Washington).

NAS/NRC (1980). National Academy of Sciences/National Research Council. *The Effects on Populations of Exposure to Low Levels of Ionizing Radiation: 1980*, Committee on the Biological Effects of Ionizing Radiations of the National Academy of Sciences/National Research Council (National Academy Press, Washington).

NCRP (1966). National Council on Radiation Protection and Measurements. *Radiation Protection in Educational Institutions*, NCRP Report No. 32 (National Council on Radiation Protection and Measurements, Bethesda, Maryland).

NCRP (1971a). National Council on Radiation Protection and Measurements. *Basic Radiation Protection Criteria*, NCRP Report No. 39 (National Council on Radiation Protection and Measurements, Bethesda, Maryland).

NCRP (1971b). National Council on Radiation Protection and Measurements. *Protection Against Neutron Radiation*, NCRP Report No. 38 (National Council on Radiation Protection and Measurements, Bethesda, Maryland).

NCRP (1977a). National Council on Radiation Protection and Measurements. *Review of NCRP Radiation Dose Limit for Embryo and Fetus in Occupationally-Exposed Women*, NCRP Report No. 53 (National Council on Radiation Protection and Measurements, Bethesda, Maryland).

NCRP (1977b). National Council on Radiation Protection and Measurements. *Medical Radiation Exposure of Pregnant and Potentially Pregnant Women*, NCRP Report No. 54 (National Council on Radiation Protection and Measurements, Bethesda, Maryland).

NCRP (1978). National Council on Radiation Protection and Measurements. *Operational Radiation Safety Program*, NCRP Report No. 59 (National Council on Radiation Protection and Measurements, Bethesda, Maryland).

NCRP (1980). National Council on Radiation Protection and Measurements. *NCRP Statement on Dose Limit for Neutrons* (National Council on Radiation Protection and Measurements, Bethesda, Maryland).

NCRP (1984a). National Council on Radiation Protection and Measurements. *Exposures from the Uranium Series with Emphasis on Radon and its Daughters*, NCRP Report No. 77 (National Council on Radiation Protection and Measurements, Bethesda, Maryland).

NCRP (1984b). National Council on Radiation Protection and Measurements. *Evaluation of Occupational and Environmental Exposure to Radon and Radon Daughters in the United States*, NCRP Report No. 78 (National Council on Radiation Protection and Measurements, Bethesda, Maryland).

NCRP (1984c). National Council on Radiation Protection and Measurements. *Control of Air Emissions of Radionuclides*, NCRP Statement (National Council on Radiation Protection and Measurements, Bethesda, Maryland).

NCRP (1985). National Council on Radiation Protection and Measurements. *General Concepts for Dosimetry of Internally Deposited Radionuclides*, NCRP Report 84 (National Council on Radiation Protection and Measurements, Bethesda, Maryland).

NCRP SC M-4 (1967). Scientific Committee M-4 of the National Council on Radiation Protection and Measurements. *Dose-Effect Modifying Factors in*

Radiation Protection, BNL 50073 (T-471) (Brookhaven National Laboratory, Upton, New York).

NIH (1985). National Institutes of Health. *Report of the National Institutes of Health Ad Hoc Working Group to Develop Radioepidemiology Tables*, NIH Publication No. 85-2748 (U.S. Government Printing Office, Washington).

NSC (1985). National Safety Council. *Accident Facts 1984* (National Safety Council, Chicago).

OTAKE, M. AND SCHULL, W. J. (1984). "In utero exposure to A-bomb radiation and mental retardation," Brit. J. Radiol. **57**, 409.

PLISKIN, J. S., SHEPARD, D. S. AND WEINSTEIN, M. C. (1980). "Utility functions for life years and health status," Operations Research **28**, 206.

POCHIN, E. E. (1981). "A perspective on risk," in *Health and Risk Analysis, Proceedings of the Third Oak Ridge National Laboratory's Life Sciences Symposium* (Franklin Institute Press, Philadelphia, Pennsylvania).

SCHOTTENFELD, D. AND HAAS, J. F. (1978). "The workplace as a cause of cancer," Clin. Bull. **8**, 54.

SINCLAIR, W. K. (1983). "Fifty years of neutrons in biology and medicine: The comparative effects of neutrons in biological systems," in Proceedings of the 8th Symposium on Microdosimetry, Julich, Germany, Sept. 27–Oct. 1, 1982. Booz, J. and Ebert, H. (Eds.), EUR-8395 en. (Office for Official Publications of European Communities, Brussels).

SINCLAIR, W. K. (1985a). "The implications of risk information for the NCRP program," page 223 in *Proceedings of the 20th Annual Meeting of the National Council on Radiation Protection and Measurements*, Proceedings No. 6 (National Council on Radiation Protection and Measurements, Bethesda, Maryland).

SINCLAIR, W. K. (1985b). "Experimental RBE values of high-LET radiations at low doses and the implications for quality factor assignment," Radiat. Prot. Dosimetry **13**, 319.

UNSCEAR (1972). United Nations Scientific Committee on the Effects of Atomic Radiation. *Ionizing Radiation: Levels and Effects*, Report to the General Assembly, with annexes, Publ. E.72.IX.18 (United Nations, New York).

UNSCEAR (1977). United Nations Scientific Committee on the Effects of Atomic Radiation. *Sources and Effects of Ionizing Radiation*, Report to the General Assembly, with annexes, Publ. E.77.IX1 (United Nations, New York).

UNSCEAR (1982). United Nations Scientific Committee on the Effects of Atomic Radiation. *Ionizing Radiation: Sources and Biological Effects*, 1982 Report to the General Assembly, with annexes, Publ. E.82.IX8 (United Nations, New York).

UNSCEAR (1986). United Nations Scientific Committee on the Effects of Atomic Radiation. *Genetic and Somatic Effects of Ionizing Radiation*, United Nations Scientific Committee on the Effects of Atomic Radiation, 1986 Report to the General Assembly, with annexes (United Nations, New York).

WILSON, R. AND CROUCH, E. A. C. (1982). *Risk/Benefit Analysis* (Ballinger Publishing Co., Cambridge, Massachusetts).

The NCRP

The National Council on Radiation Protection and Measurements is a nonprofit corporation chartered by Congress in 1964 to:

1. Collect, analyze, develop, and disseminate in the public interest information and recommendations about (a) protection against radiation and (b) radiation measurements, quantities, and units, particularly those concerned with radiation protection;
2. Provide a means by which organizations concerned with the scientific and related aspects of radiation protection and of radiation quantities, units, and measurements may cooperate for effective utilization of their combined resources, and to stimulate the work of such organizations.
3. Develop basic concepts about radiation quantities, units, and measurements, about the application of these concepts, and about radiation protection;
4. Cooperate with the International Commission on Radiological Protection, the International Commission on Radiation Units and Measurements, and other national and international organizations, governmental and private, concerned with radiation quantities, units, and measurements and with radiation protection.

The Council is the successor to the unincorporated association of scientists known as the National Committee on Radiation Protection and Measurements and was formed to carry on the work begun by the Committee.

The Council is made up of the members and the participants who serve on the eighty-two scientific committees of the Council. The scientific committees, composed of experts having detailed knowledge and competence in the particular area of the committee's interest, draft proposed recommendations. These are then submitted to the full membership of the Council for careful review and approval before being published.

The following comprise the current officers and membership of the Council:

Officers

President	Warren K. Sinclair
Vice President	S. James Adelstein
Secretary and Treasurer	W. Roger Ney
Assistant Secretary	Carl D. Hobelman
Assistant Treasurer	James F. Berg

Members

Seymour Abrahamson
S. James Adelstein
Peter R. Almond
Edward L. Alpen
John A. Auxier
William J. Bair
Michael A. Bender
John D. Boice, Jr.
Bruce B. Boecker
Robert L. Brent
Antone Brooks
Thomas F. Budinger
Melvin W. Carter
George W. Casarett
Randall S. Caswell
Fred T. Cross
Stanley B. Curtis
Gerald D. Dodd
Patricia W. Durbin
Joe A. Elder
Thomas S. Ely
Edward R. Epp
Jacob I. Fabrikant
R. J. Michael Fry
Ethyl S. Gibert
Robert A. Goepp
Joel E. Gray
Arthur W. Guy
Eric J. Hall
Naomi H. Harley
William R. Hendee
Donald G. Jacobs
A. Everette James, Jr.
Bernd Kahn
Kenneth R. Kase
Charles E. Land
George R. Leopold
Ray D. Lloyd
Arthur C. Lucas
Charles W. Mays
Roger O. McClellen
James McLaughlin
Barbara J. McNeil
Thomas F. Meaney
Charles B. Meinhold
Mortimer L. Mendelsohn
Fred A. Mettler
William E. Mills
Dade W. Moeller
A. Alan Moghissi
Robert D. Moseley, Jr.
Wesley Nyborg
Mary Ellen O'Connor
Andrew K. Poznanski
Norman C. Rasmussen
Wiliam C. Reinig
Chester R. Richmond
James S. Robertson
Lawrence N. Rothenberg
Leonard A. Sagan
William J. Schull
Glenn E. Sheline
Roy E. Shore
Warren K. Sinclair
Paul Slovic
Lewis V. Spencer
William L. Templeton
J. W. Thiessen
Roy C. Thompson
John E. Till
Arthur C. Upton
George L. Voelz
Edward W. Webster
George M. Wilkening
H. Rodney Withers
Maruin Ziskin

Honorary Members

Lauriston S. Taylor, *Honorary President*

Edgar C. Barnes
Victor P. Bond
Reynold F. Brown
Austin M. Brues
George W. Casarret
Frederick P. Cowan
James F. Crow
Merrill Eisenbud
Robley D. Evans
Richard F. Foster
Hymer L. Friedell
Robert O. Gorson
John H. Harley
John W. Healy
Louis H. Hempelmann, Jr.
Paul C. Hodges
George V. LeRoy
Wilfred B. Mann
Karl Z. Morgan
Robert J. Nelsen
Harald H. Rossi
William G. Russell
John H. Rust
Eugene L. Saenger
J. Newell Stannard
Harold O. Wyckoff

Currently, the following subgroups are actively engaged in formulating recommendations:

SC-1: Basic Radiation Protection Criteria
SC-3: Medical X-Ray, Electron Beam and Gamma-Ray Protection for Energies Up to 50 MeV (Equipment Performance and Use)
SC-16: X-Ray Protection in Dental Offices
SC-18: Standards and Measurements of Radioactivity for Radiological Use
SC-28: Radiation Exposure from Consumer Products
SC-40: Biological Aspects of Radiation Protection Criteria
Task Group on Atomic Bomb Survivor Dosimetry
Subgroup on Biological Aspects of Dosimetry of Atomic Bomb Survivors
SC-43: Natural Background Radiation
SC-44: Radiation Associated with Medical Examination
SC-45: Radiation Received by Radiation Employees
SC-46: Operational Radiation Safety
Task Group 2 on Uranium Mining and Milling—Radiation Safety Programs
Task Group 3 on ALARA for Occupationally Exposed Individuals in Clinical Radiology
Task Group 4 on Calibration of Instrumentation
Task Group 5 on Maintaining Radiation Protection Records
Task Group 6 on Radiation Protection for Allied Health Personnel
Task Group 7 on Emergency Planning
Task Group 8 on Radiation Protection Design Guidelines for Particle Accelerators
Task Group 9 on ALARA at Nuclear Power Plants
SC-47: Instrumentation for the Determination of Dose Equivalent
SC-48: Assessment of Exposure of the Population
SC-52: Conceptual Basis of Calculations of Dose Distributions
SC-57: Internal Emitter Standards
Task Group 2 on Respiratory Tract Model
Task Group 5 on Gastrointestinal Models
Task Group 6 on Bone Problems
Task Group 8 on Leukemia Risk
Task Group 9 on Lung Cancer Risk
Task Group 10 on Liver Cancer Risk
Task Group 11 on Genetic Risk
Task Group 12 on Strontium
Task Group 13 on Neptunium
Task Group 14 on Placental Transfer
Task Group 15 on Uranium
SC-59: Human Radiation Exposure Experience
SC-61: Radon Measurements
SC-63: Radiation Exposure Control in a Nuclear Emergency
Task Group on Public Knowledge About Radiation
Task Group on Exposure Criteria for Specialized Categories of the Public
SC-64: Radionuclides in the Environment
Task Group 5 on Public Exposure from Nuclear Power
Task Group 6 on Screening Models
Task Group 7 on Contaminated Soil as a Source of Radiation Exposure

Task Group 8 on Ocean Disposal of Radioactive Waste
Task Group 9 on Biological Effects on Aquatic Organisms
Task Group 10 on Low Level Waste
Task Group 11 on Xenon

SC-65: Quality Assurance and Accuracy in Radiation Protection Measurements
SC-66: Biological Effects and Exposure Criteria for Ultrasound
SC-67: Biological Effects of Magnetic Fields
SC-68: Microprocessors in Dosimetry
SC-69: Efficacy of Radiographic Procedures
SC-70: Quality Assurance and Measurement in Diagnostic Radiology
SC-71: Radiation Exposure and Potentially Related Injury
SC-74: Radiation Received in the Decontamination of Nuclear Facilities
SC-75: Guidance on Radiation Received in Space Activities
SC-76: Effects of Radiation on the Embryo-Fetus
SC-77: Guidance on Occupational and Public Exposure Resulting from Diagnostic Nuclear Medicine Procedures
SC-78: Practical Guidance on the Evaluation of Human Exposures in Radiofrequency Radiation
SC-79: Extremely Low-Frequency Electric and Magnetic Fields
SC-80: Radiation Biology of the Skin (Beta-Ray Dosimetry)
SC-81: Assessment of Exposure from Therapy
SC-82: Control of Indoor Radon

Study Group on Comparative Risk
Task Group on Comparative Carcinogenicity of Pollutant Chemicals
Ad Hoc Group on Medical Evaluation of Radiation Workers
Ad Hoc Group on Video Display Terminals
Task Force on Occupational Exposure Levels

In recognition of its responsibility to facilitate and stimulate cooperation among organizations concerned with the scientific and related aspects of radiation protection and measurement, the Council has created a category of NCRP Collaborating Organizations. Organizations or groups of organizations that are national or international in scope and are concerned with scientific problems involving radiation quantities, units, measurements, and effects, or radiation protection may be admitted to collaborating status by the Council. The present Collaborating Organizations with which the NCRP maintains liaison are as follows:

American Academy of Dermatology
American Association of Physicists in Medicine
American College of Nuclear Physicians
American College of Radiology
American Dental Association
American Industrial Hygiene Association
American Institute of Ultrasound in Medicine
American Insurance Association
American Medical Association
American Nuclear Society

American Occupational Medical Association
American Podiatric Medical Association
American Public Health Association
American Radium Society
American Roentgen Ray Society
American Society of Radiologic Technologists
American Society for Therapeutic Radiology and Oncology
Association of University Radiologists
Atomic Industrial Forum
Bioelectromagnetics Society
College of American Pathologists
Conference on Radiation Control Program Directors
Federal Communications Commission
Federal Emergency Management Agency
Genetics Society of America
Health Physics Society
National Bureau of Standards
National Electrical Manufacturers Association
Radiation Research Society
Radiological Society of North America
Society of Nuclear Medicine
United States Air Force
United States Army
United States Department of Energy
United States Department of Housing and Urban Development
United States Department of Labor
United States Environmental Protection Agency
United States Navy
United States Nuclear Regulatory Commission
United States Public Health Service

The NCRP has found its relationships with these organizations to be extremely valuable to continued progress in its program.

Another aspect of the cooperative efforts of the NCRP relates to the special liaison relationships established with various governmental organizations that have an interest in radiation protection and measurements. This liaison relationship provides: (1) an opportunity for participating organizations to designate an individual to provide liaison between the organization and the NCRP; (2) that the individual designated will receive copies of draft NCRP reports (at the time that these are submitted to the members of the Council) with an invitation to comment, but not vote; and (3) that new NCRP efforts might be discussed with liaison individuals as appropriate, so that they might have an opportunity to make suggestions on new studies and related matters. The following organizations participate in the special liaison program:

Commission of the European Communities
Commisariat a l'Energie Atomique (France)

Defense Nuclear Agency
Federal Emergency Management Agency
Japan Radiation Council
National Bureau of Standards
National Radiological Protection Board (United Kingdom)
National Research Council (Canada)
Office of Science and Technology Policy
Office of Technology Assessment
United States Air Force
United States Army
United States Coast Guard
United States Department of Energy
United States Department of Health and Human Services
United States Department of Labor
United States Department of Transportation
United States Environmental Protection Agency
United States Navy
United States Nuclear Regulatory Commission

The NCRP values highly the participation of these organizations in the liaison program.

The Council's activities are made possible by the voluntary contribution of time and effort by its members and participants and the generous support of the following organizations:

Alfred P. Sloan Foundation
Alliance of American Insurers
American Academy of Dental Radiology
American Academy of Dermatology
American Association of Physicists in Medicine
American College of Nuclear Physicians
American College of Radiology
American College of Radiology Foundation
American Dental Association
American Hospital Radiology Administrators
American Industrial Hygiene Association
American Insurance Association
American Medical Association
American Nuclear Society
American Occupational Medical Association
American Osteopathic College of Radiology
American Podiatric Medical Association
American Public Health Association
American Radium Society
American Roentgen Ray Society
American Society of Radiologic Technologists
American Society for Therapeutic Radiology and Oncology
American Veterinary Medical Association
American Veterinary Radiology Society
Association of University Radiologists

Atomic Industrial Forum
Battelle Memorial Institute
Center for Devices and Radiological Health
College of American Pathologists
Commonwealth of Pennsylvania
Defense Nuclear Agency
Edison Electric Institute
Edward Mallinckrodt, Jr. Foundation
Electric Power Research Institute
Federal Emergency Management Agency
Florida Institute of Phosphate Research
Genetics Society of America
Health Physics Society
Institute of Nuclear Power Operations
James Picker Foundation
National Aeronautics and Space Administration
National Association of Photographic Manufacturers
National Bureau of Standards
National Cancer Institute
National Electrical Manufacturers Association
Radiation Research Society
Radiological Society of North America
Society of Nuclear Medicine
United States Department of Energy
United States Department of Labor
United States Environmental Protection Agency
United States Navy
United States Nuclear Regulatory Commission

To all of these organizations the Council expresses its profound appreciation for their support.

Initial funds for publication of NCRP reports were provided by a grant from the James Picker Foundation and for this the Council wishes to express its deep appreciation.

The NCRP seeks to promulgate information and recommendations based on leading scientific judgment on matters of radiation protection and measurement and to foster cooperation among organizations concerned with these matters. These efforts are intended to serve the public interest and the Council welcomes comments and suggestions on its reports or activities from those interested in its work.

NCRP Publications

NCRP publications are distributed by the NCRP Publications' office. Information on prices and how to order may be obtained by directing an inquiry to:

NCRP Publications
7910 Woodmont Ave., Suite 1016
Bethesda, MD 20814

The currently available publications are listed below.

Proceedings of the Annual Meeting

No.	Title
1	*Perceptions of Risk*, Proceedings of the Fifteenth Annual Meeting, Held on March 14–15, 1979 (Including Taylor Lecture No. 3) (1980)
2	*Quantitative Risk in Standards Setting*, Proceedings of the Sixteenth Annual Meeting, Held on April 2–3, 1980 (Including Taylor Lecture No. 4) (1981)
3	*Critical Issues in Setting Radiation Dose Limits*, Proceedings of the Seventeenth Annual Meeting, Held on April 8–9, 1981 (Including Taylor Lecture No. 5) (1982)
4	*Radiation Protection and New Medical Diagnostic Procedures*, Proceedings of the Eighteenth Annual Meeting, Held on April 6–7, 1982 (Including Taylor Lecture No. 6) (1983)
5	*Environmental Radioactivity*, Proceedings of the Nineteenth Annual Meeting, Held on April 6–7, 1983 (Including Taylor Lecture No. 7) (1984)
6	*Some Issues Important in Developing Basic Radiation Protection Recommendations*, Proceedings of the Twentieth Annual Meeting, Held on April 4–5, 1984 (Including Taylor Lecture No. 8) (1985)
7	*Radioactive Waste*, Proceedings of the Twenty-First Annual Meeting, Held on April 3–4, 1985 (Including Taylor Lecture No. 9) (1986)

Symposium Proceedings

The Control of Exposure of the Public to Ionizing Radiation in the Event of Accident or Attack, Proceedings of a Symposium held April 27–29, 1981 (1982)

Lauriston S. Taylor Lectures

No.	Title and Author
1	*The Squares of the Natural Numbers in Radiation Protection* by Herbert M. Parker (1977)
2	*Why be Quantitative About Radiation Risk Estimates?* by Sir Edward Pochin (1978)
3	*Radiation Protection—Concepts and Trade Offs* by Hymer L. Friedell (1979) [Available also in *Perceptions of Risk*, see above]
4	*From "Quantity of Radiation" and "Dose" to "Exposure" and "Absorbed Dose"—An Historical Review* by Harold O. Wyckoff (1980) [Available also in *Quantitative Risks in Standards Setting*, see above]
5	*How Well Can We Assess Genetic Risk? Not Very* by James F. Crow (1981) [Available also in *Critical Issues in Setting Radiation Dose Limits*, see above]
6	*Ethics, Trade-offs and Medical Radiation* by Eugene L. Saenger (1982) [Available also in *Radiation Protection and New Medical Diagnostic Approaches*, see above]
7	*The Human Environment-Past, Present and Future* by Merril Eisenbud (1983) [Available also in *Environmental Radioactivity*, see above]
8	*Limitation and Assessment in Radiation Protection* by Harald H. Rossi (1984) [Available also in *Some Issues Important in Developing Basic Radiation Protection Recommendations*, see above]
9	*Truth (and Beauty) in Radiation Measurement* by John H. Harley (1985)
10	*Nonionizing Radiation Bioeffects: Cellular Properties and Interactions* by Herman P. Schwan (1986)

NCRP Commentaries

Commentary No.	Title
1	*Krypton-85 in the Atmosphere—With Specific Reference to the Public Health Significance of the Proposed Controled Release at Three Mile Island* (1980)

2 *Preliminary Evaluation of Criteria for the Disposal of Transuranic Contaminated Waste* (1982)

3 *Screening Techniques for Determining Compliance with Environmental Standards* (1986)

4 *Guidelines for the Release of Waste Water from Nuclear Facilities with Special Reference to the Public Health Significance of the Proposed Release of Treated Waste Waters at Three Mile Island*

NCRP Reports

No.	Title
8	*Control and Removal of Radioactive Contamination in Laboratories* (1951)
9	*Recommendations for Waste Disposal of Phosphorus-32 and Iodine-131 for Medical Users* (1951)
16	*Radioactive Waste Disposal in the Ocean* (1954)
22	*Maximum Permissible Body Burdens and Maximum Permissible Concentrations of Radionuclides in Air and in Water for Occupational Exposure* (1959) [Includes Addendum 1 issued in August 1963]
23	*Measurement of Neutron Flux and Spectra for Physical and Biological Applications* (1960)
25	*Measurement of Absorbed Dose of Neutrons and Mixtures of Neutrons and Gamma Rays* (1961)
27	*Stopping Powers for Use with Cavity Chambers* (1961)
30	*Safe Handling of Radioactive Materials* (1964)
32	*Radiation Protection in Educational Institutions* (1966)
33	*Medical X-Ray and Gamma-Ray Protection for Energies Up to 10 MeV—Equipment Design and Use* (1968)
35	*Dental X-Ray Protection* (1970)
36	*Radiation Protection in Veterinary Medicine* (1970)
37	*Precautions in the Management of Patients Who Have Received Therapeutic Amounts of Radionuclides* (1970)
38	*Protection against Neutron Radiation* (1971)
40	*Protection Against Radiation from Brachytherapy Sources* (1972)
41	*Specification of Gamma-Ray Brachytherapy Sources* (1974)
42	*Radiological Factors Affecting Decision-Making in a Nuclear Attack* (1974)
44	*Krypton-85 in the Atmosphere—Accumulation, Biological Significance, and Control Technology* (1975)
45	*Natural Background Radiation in the United States* (1975)

46 *Alpha-Emitting Particles in Lungs* (1975)
47 *Tritium Measurement Techniques* (1976)
48 *Radiation Protection for Medical and Allied Health Personnel* (1976)
49 *Structural Shielding Design and Evaluation for Medical Use of X Rays and Gamma Rays of Energies Up to 10 MeV* (1976)
50 *Environmental Radiation Measurements* (1976)
51 *Radiation Protection Design Guidelines for 0.1–100 MeV Particle Accelerator Facilities* (1977)
52 *Cesium-137 From the Environment to Man: Metabolism and Dose* (1977)
53 *Review of NCRP Radiation Dose Limit for Embryo and Fetus in Occupationally Exposed Women* (1977)
54 *Medical Radiation Exposure of Pregnant and Potentially Pregnant Women* (1977)
55 *Protection of the Thyroid Gland in the Event of Releases of Radioiodine* (1977)
56 *Radiation Exposure From Consumer Products and Miscellaneous Sources* (1977)
57 *Instrumentation and Monitoring Methods for Radiation Protection* (1978)
58 *A Handbook of Radioactivity Measurements Procedures*, 2nd ed. (1985)
59 *Operational Radiation Safety Program* (1978)
60 *Physical, Chemical, and Biological Properties of Radiocerium Relevant to Radiation Protection Guidelines* (1978)
61 *Radiation Safety Training Criteria for Industrial Radiography* (1978)
62 *Tritium in the Environment* (1979)
63 *Tritium and Other Radionuclide Labeled Organic Compounds Incorporated in Genetic Material* (1979)
64 *Influence of Dose and Its Distribution in Time on Dose-Response Relationships for Low-LET Radiations* (1980)
65 *Management of Persons Accidentally Contaminated with Radionuclides* (1980)
66 *Mammography* (1980)
67 *Radiofrequency Electromagnetic Fields—Properties, Quantities and Units, Biophysical Interaction, and Measurements* (1981)
68 *Radiation Protection in Pediatric Radiology* (1981)
69 *Dosimetry of X-Ray and Gamma-Ray Beams for Radiation Therapy in the Energy Range 10 keV to 50 MeV* (1981)
70 *Nuclear Medicine-Factors Influencing the Choice and Use*

of Radionuclides in Diagnosis and Therapy (1982)

71 *Operational Radiation Safety—Training* (1983)

72 *Radiation Protection and Measurement for Low Voltage Neutron Generators* (1983)

73 *Protection in Nuclear Medicine and Ultrasound Diagnostic Procedures in Children* (1983)

74 *Biological Effects of Ultrasound: Mechanisms and Clinical Implications* (1983)

75 *Iodine-129: Evaluation of Releases from Nuclear Power Generation* (1983)

76 *Radiological Assessment: Predicting the Transport, Bioaccumulation, and Uptake by Man of Radionuclides Released to the Environment* (1984)

77 *Exposures from the Uranium Series with Emphasis on Radon and its Daughters* (1984)

78 *Evaluation of Occupational and Environmental Exposures to Radon and Radon Daughters in the United States* (1984)

79 *Neutron Contamination from Medical Electron Accelerators* (1984)

80 *Induction of Thyroid Cancer by Ionizing Radiation* (1985)

81 *Carbon-14 in the Environment* (1985)

82 *SI Units in Radiation Protection and Measurements* (1985)

83 *The Experimental Basis for Absorbed Dose-Calculations in Medical uses of Radionuclides* (1985)

84 *General Concepts for the Dosimetry of Internally Deposited Radionuclides* (1985)

85 *Mammography—A User's Guide* (1986)

86 *Biological Effects and Exposure Criteria for Radiofrequency Electromagnetic Fields* (1986)

87 *Use of Bioassay Procedures for Assessment of Internal Radionuclide Deposition*

88 *Radiation Alarms and Access–Control Systems* (1987)

89 *Genetic Effects of Internally Deposited Radionuclides* (1987)

90 *Neptunium: Radiation Protection Guidelines* (1987)

91 *Recommendations on Limits for Exposure to Ionizing Radiation* (1987)

Binders for NCRP Reports are available. Two sizes make it possible to collect into small binders the "old series" of reports (NCRP Reports Nos. 8–30) and into large binders the more recent publications (NCRP Reports Nos. 32–88). Each binder will accommodate from five to seven reports. The binders carry the identification "NCRP Reports" and come with label holders which permit the user to attach labels showing the reports contained in each binder.

The following bound sets of NCRP Reports are also available:

Volume I. NCRP Reports Nos. 8, 9, 12, 16, 22
Volume II. NCRP Reports Nos. 23, 25, 27, 30
Volume III. NCRP Reports Nos. 32, 33, 35, 36, 37
Volume IV. NCRP Reports Nos. 38, 39, 40, 41
Volume V. NCRP Reports Nos. 42, 43, 44, 45, 46
Volume VI. NCRP Reports Nos. 47, 48, 49, 50, 51
Volume VII. NCRP Reports Nos. 52, 53, 54, 55, 56, 57
Volume VIII. NCRP Reports No. 58
Volume IX. NCRP Reports Nos. 59, 60, 61, 62, 63
Volume X. NCRP Reports Nos. 64, 65, 66, 67
Volume XI. NCRP Reports Nos. 68, 69, 70, 71, 72
Volume XII. NCRP Reports Nos. 73, 74, 75, 76
Volume XIII. NCRP Reports Nos. 77, 78, 79, 80
Volume XIV. NCRP Reports Nos. 81, 82, 83, 84, 85.

Titles of the individual reports contained in each volume are given above).

The following NCRP Reports are now superseded and/or out of print:

No.	Title
1	*X-Ray Protection* (1931). [Superseded by NCRP Report No. 3]
2	*Radium Protection* (1934). [Superseded by NCRP Report No. 4]
3	*X-Ray Protection* (1936). [Superseded by NCRP Report No. 6]
4	*Radium Protection* (1938). [Superseded by NCRP Report No. 13]
5	*Safe Handling of Radioactive Luminous Compounds* (1941). [Out of Print]
6	*Medical X-Ray Protection Up to Two Million Volts* (1949). [Superseded by NCRP Report No. 18]
7	*Safe Handling of Radioactive Isotopes* (1949). [Superseded by NCRP Report No. 30]
10	*Radiological Monitoring Methods and Instruments* (1952). [Superseded by NCRP Report No. 57]
11	*Maximum Permissible Amounts of Radioisotopes in the Human Body and Maximum Permissible Concentrations in Air and Water* (1953). [Superseded by NCRP Report No. 22]

12 *Recommendations for the Disposal of Carbon-14 Wastes* (1953). [Superseded by NCRP Report No. 81]

13 *Protection Against Radiations from Radium, Cobalt-60 and Cesium-137* (1954). [Superseded by NCRP Report No. 24]

14 *Protection Against Betatron—Synchrotron Radiations Up to 100 Million Electron Volts* (1954). [Superseded by NCRP Report No. 51]

15 *Safe Handling of Cadavers Containing Radioactive Isotopes* (1953). [Superseded by NCRP Report No. 21]

17 *Permissible Dose from External Sources of Ionizing Radiation* (1954) including *Maximum Permissible Exposure to Man, Addendum to National Bureau of Standards Handbook 59* (1958). [Superseded by NCRP Report No. 39]

18 *X-Ray Protection* (1955). [Superseded by NCRP Report No. 26]

19 *Regulation of Radiation Exposure by Legislative Means* (1955). [Out of Print]

20 *Protection Against Neutron Radiation Up to 30 Million Electron Volts* (1957). [Superseded by NCRP Report No. 38]

21 *Safe Handling of Bodies Containing Radioactive Isotopes* (1958). [Superseded by NCRP Report No. 37]

24 *Protection Against Radiations from Sealed Gamma Sources* (1960). [Superseded by NCRP Report Nos. 33, 34, and 40]

26 *Medical X-Ray Protection Up to Three Million Volts* (1961). [Superseded by NCRP Report Nos. 33, 34, 35, and 36]

28 *A Manual of Radioactivity Procedures* (1961). [Superseded by NCRP Report No. 58]

29 *Exposure to Radiation in an Emergency* (1962). [Superseded by NCRP Report No. 42]

31 *Shielding for High Energy Electron Accelerator Installations* (1964). [Superseded by NCRP Report No. 51]

34 *Medical X-Ray and Gamma-Ray Protection for Energies Up to 10 MeV—Structural Shielding Design and Evaluation* (1970). [Superseded by NCRP Report No. 49]

39 *Basic Radiation Protection Criteria* (1971). [Superseded by NCRP Report No. 91]

43 *Review of the Current State of Radiation Protection Philosophy* (1975). [Superseded by NCRP Report No. 91]

Other Documents

The following documents of the NCRP were published outside of the NCRP Reports series:

"Blood Counts, Statement of the National Committee on Radiation Protection," Radiology 63, 428 (1954)

"Statements on Maximum Permissible Dose from Television Receivers and Maximum Permissible Dose to the Skin of the Whole Body," Am. J. Roentgenol., Radium Ther. and Nucl. Med. 84, 152 (1960) and Radiology 75, 122 (1960)

X-Ray Protection Standards for Home Television Receivers, Interim Statement of the National Council on Radiation Protection and Measurements (National Council on Radiation Protection and Measurements, Washington, 1968)

Specification of Units of Natural Uranium and Natural Thorium (National Council on Radiation Protection and Measurements, Washington, 1973)

NCRP Statement on Dose Limit for Neutrons (National Council on Radiation Protection and Measurements, Washington, 1980)

Control of Air Emissions of Radionuclides (National Council on Radiation Protection and Measurements, Bethesda, Maryland, 1984)

Copies of the statements published in journals may be consulted in libraries. A limited number of copies of the remaining documents listed above are available for distribution by NCRP Publications.

INDEX